About the authors

Patrick Holford is a leading light in new
approaches to health and nutrition. He began
his academic career in the field of psychology.
While completing his bachelor degree in
Experimental Psychology at the University of
York he researched the role of nutrition in
mental health and illness and later tested the
effects of improved nutrition on children's IQ –
an experiment that was the subject of a *Horizon*
documentary in 1987. He became a student of
twice Nobel Prize winner Dr Linus Pauling,

who believed that the future of medicine was 'optimum nutrition'.

In 1984, with the support of Dr Pauling, Patrick Holford founded the
Institute for Optimum Nutrition. A charitable and independent
educational trust for furthering education and research in nutrition,
ION is now one of the most respected training colleges for clinical
nutritionists. At ION he pioneered many radical ideas in nutrition, from
the importance of antioxidants to the dangers of HRT.

He has written 20 popular books, now translated into 17 languages.
The first, *The Optimum Nutrition Bible*, has sold over a million copies
worldwide. Backed by more than 20 years of research and clinical
experience, Patrick Holford is convinced that homocysteine is the next
big breakthrough in preventative medicine.

Dr James Braly qualified in medicine in 1970
and has remained at the cutting edge of new
American medical research in natural medicine
and laboratory science for the past 23 years.

Disenchanted with prescription drugs and
symptom-oriented treatments, he founded
innovative naturopathic medical clinics in
Encino and San Mateo, California (1980–1994),
specialising in an optimum nutrition approach
to health care, backed up by proper laboratory
testing. He founded and directed one of the first

licensed clinical laboratories in the world for testing IgG-based food
intolerance. Both through his clinics and laboratory he supervised the
treatment of thousands of people using nutrition-based, non-drug
treatments for preventing and reversing disease.

He was one of the first doctors to alert us to the epidemic of IgG
food allergies and is author of the best-selling book *Dr Braly's Food*

Allergy and Nutrition Revolution, as well as the more recent book *Food Allergy Relief*. He is also a recognised world authority on coeliac disease and has helped to research and expose the widespread problem of gluten sensitivity, culminating in his recent book *Dangerous Grains*. He is medical editor and consultant for www.drbralyallergyrelief.com, a website specialising in consumer education about clinical nutrition, delayed-onset food allergy, gluten sensitivity and coeliac disease. He is currently involved as medical director, researcher, nutrition product formulator and consultant with the California-based LifeStream Recovery, Inc., dealing with innovative nutritional, neurochemical, immunological and hormonal treatments of addictions and neurodegenerative diseases.

In recent years Dr Braly has extensively researched the role of homocysteine in health and disease and is convinced that homocysteine is the single greatest predictor of health and disease. His pioneering research has also identified how to lower homocysteine rapidly, thus reducing the risk of many life-threatening and debilitating diseases.

THE
H
FACTOR

**HOMOCYSTEINE – THE BIGGEST HEALTH
BREAKTHROUGH OF THE CENTURY**

PATRICK HOLFORD
& DR JAMES BRALY MD

PIATKUS

Copyright © 2003 by Patrick Holford and Dr James Braly

First published in 2003 by
Judy Piatkus (Publishers) Limited
5 Windmill Street
London W1T 2JA
e-mail: info@piatkus.co.uk

Reprinted 2003

The moral right of the authors has been asserted

A catalogue record for this book is available from the British Library

ISBN 0 7499 2419 5

Edited by Barbara Kiser
Text design by Paul Saunders

This book has been printed on paper manufactured with respect for the environment using wood from managed sustainable resources

Data manipulation by
Phoenix Photosetting, Chatham, Kent
Printed and bound in Great Britain by
Antony Rowe Ltd, Chippenham, Wilts

Contents

Ten Steps to Superhealth

Acknowledgements

This book would not have been possible without the help, support and research of many people, especially our families for putting up with the early mornings and late nights. We would also like to thank John Graham and his team at YorkTest for their help with case histories and Susannah Lawson for her help with editing, researching and referencing. We would also like to thank our publishers Piatkus, especially Gill, Anna and Barbara on editorial, as well as Philip, Jana and Judy for their support, encouragement and enthusiasm. And finally, we would like to thank Dr Kilmer McCully for having the insight, courage and perseverance to research the homocysteine hypothesis tirelessly against all odds, and Dr Craig Cooney for his invaluable insight into the fundamental importance of methylation in health and disease.

Guide to abbreviations, measures and references

Homocysteine is measured in micromoles per litre (mmol/l). We call these 'units'.

Most vitamins are measured in milligrams or micrograms.

1 gram (g) = 1,000 milligrams (mg) = 1,000,000 micrograms (mcg)

In each part of the book, you'll find numbered references. These refer to research papers listed in the References section on page 243, and are there for readers who want to study this subject in depth.

Introduction

STAYING HEALTHY, HAPPY, clear-headed and full of energy into old age – this is what we all want. But ensuring that we do depends on how well we can 'read' the state of our health. What if there was a single test that could do that, as well as pointing the way forward to a superhealthy future?

This book is about an extraordinary discovery that has taken the world of medicine by storm. Open almost any medical journal today and you'll find something about homocysteine, a substance with a remarkable trait. Like a chemical crystal ball, it reveals exactly what we should be doing to guarantee our future health.

This amazing predictive chemical is an amino acid – a building block of protein – found naturally in the blood. Our levels of it, or our 'H score', are more accurate than a cholesterol reading in terms of predicting our risk of a heart attack or stroke, and a better measure than genes of our risk of Alzheimer's. In fact, your H score predicts your risk of over 50 diseases – including increased risk of premature death from all common causes. It can even tell you how quickly you will age. And by revealing your B vitamin nutritional status, immune system function and the state of your brain, it shows how well your body's chemistry can roll with the punches.

If this was the whole story, many of us would rather not know. But a high H score is anything but cause for gloom. The reason? It's remarkably easy to fix. In a mere two months you can ensure your future health, potentially halving your risk of most life-threatening diseases and even slowing down the ageing process. We know how, and we want you to know.

You might already be taking some steps in the right direction. Maybe you've quit smoking, lost some weight, lowered your cholesterol, or are taking a high-quality daily multivitamin. But how do you really know where you stand on the scale of health and disease? If you don't know your H score, you won't ever really know.

What you will discover in this book is that most people's H scores, which can now be tested at home using a simple kit, are around 11 points. Probably two in ten people reading this book will have a superhealthy H score below 6, and two in ten will have an extremely unhealthy H score above 15. But what does this mean?

Researchers from the University of Bergen in Norway, one of the world's leading homocysteine research centres, measured the homocysteine levels of 4,766 people in their sixties a decade ago, and then recorded who died and who didn't. This is what they discovered.

A mere five-point decrease in H score predicted a:

- 49% reduced risk of death from all causes

- 50% reduced risk of death from cardiovascular disease

- 26% reduced risk of death from cancer

- 104% reduced risk of death from any causes other than cancer or heart disease.

These are pretty staggering statistics, and you'll find them confirmed again and again by researchers all over the world. But let's look at what this means in real terms.

Let's say you're average, with an H score of 11 points. By following our H Factor programme, which you'll find in Parts 4 and 5 of this book, your H score should drop about five points in two months. And your risk of getting over 50 diseases will also substantially decrease. If your H score is 16 and you follow the H Factor programme for four months, you'll decrease your score by up to ten points and end up with a fraction of the risk you started with.

And we're not just talking about adding years to your life. We're also looking at how to add life to your years.

Whether you are young or old, sick or healthy, the best time to measure your H score is now. Then, if necessary, you can do what you need to do to bring it into the healthy range. If you're young, remember that homocysteine does much more than predict risk for diseases – it also shows how well we'll cope with life's long rollercoaster ride.

case study

Alan is a case in point. His father died of heart disease at the age of 56, so not surprisingly, being 56, he was concerned about the history of heart disease in his family. Alan also used to be very active but as he got older he did less exercise – and then he'd have a glass of wine, or two, a night. So he thought he'd get a checkover.

His cholesterol was normal, as was his blood pressure. His homocysteine, however, was 14.6, putting him in the high risk category. He started supplementing B6, B12 and folic acid. Within three months his homocysteine level was 9.6, reducing his risk of a heart attack by more than a third.

■ Homocysteine measures your body's age and intelligence

It's an incredible fact that not only are we literally made from the food we eat, but that 70 per cent of our bodies are also renewed every year. The 'you' that stares back from the bathroom mirror isn't the same 'you' that stared back a year ago! It is, quite literally, a miracle. So, what is the 'intelligence' in our make-up that keeps making us new, or nearly new, every year?

It's all down to two extraordinary chemical balancing acts in the body. One is called methylation. The other is oxidation-reduction. If you can get these two working at peak levels you are going to feel great, and stay vibrantly young – mentally, emotionally and physically. Both are measured and predicted by your blood levels of homocysteine and both can be improved by the right nutrients, gleaned from both diet and supplements.

And this is a vital step. Methylation and oxidation-reduction largely control how rapidly you age and how early chronic disease sets in. If you lack antioxidants in your diet, such as vitamins A, C, E and selenium, you age more quickly and get sick sooner. The same happens if you lack 'methylating agents' in your diet, such as vitamin B12 and folic acid. These two chemical processes have got a lot more to do with your lifespan and pattern of diseases than your genes.

In fact, even gene 'mutations' that increase the risk of disease can be controlled by methylation. According to Dr Adrian Bird of Edinburgh University, 'One in three mutations that cause human disease can be attributed to methyl groups on our genes.'

Homocysteine reflects the health of your genes, how well you are holding back the clock and your risk of premature death from all common causes. That's how all-encompassing the homocysteine story is. Your H score is more important than your weight, your blood pressure or your cholesterol level. Quite simply, it is your most vital, preventable and reversible health statistic.

▪ Lowering homocysteine is easy with the H Factor programme

So, let's say your H score is higher than you'd like. What can you do over the next two months? All the H Factor programme involves are some simple changes to your diet, and a daily homocysteine-busting supplement or two. For a few pence extra a day, you can progress from potential health risk to maximising your health potential.

Using homocysteine as our yardstick, we've been researching exactly what kind of diet, supplement and lifestyle put you in the 'superhealth' category. We've looked at meat versus vegetables, wine versus beer, the effects of coffee, smoking, salt, being over-weight and many other potentially harmful or protective foods, drinks and habits. And there are some surprises in store.

As you might expect, excess alcohol and cigarettes are extremely bad news. But do you have to quit completely? Yes, for smoking. No, for alcohol. In fact, small amounts of alcohol may even be beneficial. Smokers, on the other hand, have high H scores which plummet on stopping, but barely change on cutting back. Strict vegetarians are more likely to have a high H score than meat eaters. We'll explain why. Going on a weight-loss diet increases, rather than decreases, H scores. This doesn't mean you shouldn't diet, but does mean you definitely shouldn't go on one without taking homocysteine-lowering supplements.

Speaking of supplements, the evidence is clear. Certain vita-min supplements lower your H score much more effectively than diet changes alone. We advocate both supplements and dietary change and will tell you exactly what to take depending on your H score.

As you will see, the right combination of four B vitamins, zinc and a nutrient you might never have heard of – trimethylglycine (TMG) – can lower homocysteine in a matter of weeks. There are other beneficial supplements, too. The medical profession has homed in on folic acid as the 'answer' and, both in Britain and the US, folic acid supplements are starting to be more widely recommended, both by doctors and by government health advisers. However, folic acid on its own often can't lower a high H level down to within a safe range. The right combination of nutrients, on the other hand, works almost every time.

Interestingly, there's no drug yet discovered that lowers your H level, which is probably why the medical profession, used to massive campaigns to launch new drugs, has been slow to embrace this vital breakthrough. With no drug, there's no great profit to be made. In fact, exactly the opposite. If you do what we recommend in this book, you could bankrupt a few of the pharmaceutical companies who are selling drugs that treat only symptoms of disease, while promoting companies and therapies that treat underlying causes of diseases with the right diet, lifestyle and supplements.

We've put all this together into an easy, economical plan. All you do is measure your H score to know exactly where you are right now, then follow the recommendations in Parts 4 and 5, retest and see your H score fall. Once you hit the superhealthy goal – an H score below 6 – you've programmed yourself for superhealth.

About this book

One last note before we plunge in to the particulars. We have made some bold claims here and in the rest of the book we'll show you how everything we say is backed up with good science – lots of it!

- **Part 1** explains what homocysteine is, why it's so bad when elevated, how lowering it decreases your risk of having the top five killer diseases, and how widespread the homocysteine problem is the world over.

- **Part 2** explains the risk factors for high homocysteine, and how you can test your own H level and see where you are on the scale from health to disease.

- **Part 3** gives you detailed information on some of the 50 medical conditions now linked to high homocysteine. If you suffer from any of these diseases, we want you to know the whole story.

- **Part 4** gives you the *raison d'être* of the H Factor programme. And it shows you why too much meat or extreme vegetarian diets can be bad news, why a little (but not too much) alcohol may help, and why caffeine, smoking and stress all raise homocysteine – and what you can do about it.

- **Part 5** explains why certain supplements can lower high homocysteine levels and which combinations and doses work best, depending on your H score.

- **Ten steps to superhealth** puts it all together into a simple ten-step action plan to bring your homocysteine level into the superhealthy zone.

Wishing you the best of health,

Patrick Holford and Dr James Braly

PART 1

The Homocysteine Story

1

The health secret of homocysteine

WE OWE A LOT, and very possibly our lives, to one smart, courageous man – Dr Kilmer McCully. His discovery that homocysteine is probably your single most vital health statistic, and his tenacity in proving it despite immense resistance, has led to a revolutionary breakthrough in medicine. As this book unfolds, you will see how the discovery of homocysteine has revealed the cause of heart disease, strokes, cancer, diabetes, Alzheimer's and over 50 other common diseases, and, most importantly, how this knowledge can help you to prevent yourself from ever getting them. But let's begin at the beginning.

Way back in 1968 McCully, who trained at Harvard Medical School, was studying children with a rare genetic disorder called 'homocystinuria'. Children born with this condition lack certain enzymes required to turn a naturally occurring yet potentially toxic substance, homocysteine, into a harmless substance called cystathionine. As a consequence, they have extraordinarily high levels of homocysteine in their blood, well above 100 units. To give you some idea of how serious this is, even a score above 9 is now considered high risk. Unless they are diagnosed and treated, these children often die at a very young age of heart attacks and strokes, once thought to be caused in large part by high cholesterol, despite having completely normal cholesterol levels. One infant died at the age of two months from advanced arteriosclerosis. And sufferers often have a variety of other ailments.

When he was reviewing these cases, McCully came to realise that homocysteine, rather than cholesterol, might be a fundamental independent risk factor of arteriosclerosis and atherosclerosis, and therefore of heart attacks and strokes. And he went on to prove that it is. But the medical profession wasn't ready to listen. McCully was dismissed from Harvard and shuffled from job to job, insistent on continuing this 'unpopular' line of research, which was then sidelined in cardiovascular medicine until the 1990s. Over the past decade, thanks to McCully's hard work and tenacity, evidence for the homocysteine theory has piled up. Now, finally, mainstream medicine is sitting up and taking note.[1]

The initial research focused on the relationship of homocysteine to cardiovascular and cerebrovascular disease – heart attacks, strokes and artery disease. In 1992 a study of 14,000 male doctors found that those with homocysteine levels in the top 5 per cent had three times the risk of having a heart attack, compared to those in the bottom 5 per cent. This increased risk was confirmed in 1995 by the internationally acclaimed Massachusetts-based Framingham Heart Study, which found that having more than 11.4 micromoles of homocysteine per litre of blood (a measurement we call 'units') increased the risk significantly.[2] Another study at the University of Washington found that having a high H score doubled the risk of heart attack even in young women.

The real clincher was a study carried out by the European Concerted Action Group, a consortium of doctors and researchers from 19 medical centres in nine European countries.[3] They studied 750 people under the age of 60 with blockages in their coronary arteries, and compared them to 800 people without cardiovascular disease. They found that having a high H score was as great a risk factor for cardiovascular disease as smoking or having a high blood cholesterol level. Those whose homocysteine levels were in the top fifth doubled the risk of cardiovascular disease. Most significantly, the group also found that those taking vitamin supplements had a risk factor two-thirds less than those not taking them. That's because certain supplements lower your H score and your risk.

Today, totally vindicated, Dr McCully is a best-selling author and pathologist at the Veterans Affairs Medical Center in Providence, Rhode Island. And in conventional cardiology, the branch of medicine that deals with heart disease, homocysteine is rapidly being recognised as a significant independent cause of risk, on a par with cholesterol. (Later on we will show you why high homocysteine is in fact a far greater risk factor for heart disease than cholesterol.) However, this link to heart disease is just the tip of the iceberg: the discovery of homocysteine has identified a fundamental cause and biological indicator, or marker, of all ageing, in every single one of us.

■ Reduce your risk of the top five killer diseases

At least one in two people now die prematurely from preventable diseases, and over one in two have high homocysteine levels. Huge numbers of us tend to adopt the ostrich technique, and bury our heads in the sand, letting all that sensible health advice wash over us – until it's too late. Our clinics are full of people who've had a heart attack or stroke, or whose mental gears are slipping into a state of senility, or who have been diagnosed with cancer or diabetes. It's only then that, out of desperation, they become willing to listen to medical advice.

It doesn't have to be this way. There is a way to age slowly and gracefully, physically vital and mentally sharp to the very end of a long and healthy life. It's prevention – always much simpler and more pleasant than cure. Remember, the top five killer diseases – heart attacks, strokes, cancer, the complications of diabetes and Alzheimer's – are mostly preventable.

The trick is to start your disease prevention strategy BEFORE you have the disease. How? One of the best places to begin is to find out your H score. If your homocysteine levels are too high, you can bring them down to safe levels with our simple H Factor programme. Once you've done that, you've effectively insured yourself for a healthy future. You'll find out all about this in Parts 4 and 5.

But knowing your H score doesn't only give you an indictor of what measures you need to take to protect yourself from many diseases. It's also a clear indicator of your ability to fight infections, your mental and physical energy and how quickly you are ageing.

■ Extending your healthy lifespan

Once you hit 65, statistics say you'll live another 16 years, if you're a man, and 18 years, if you're a woman. Amazingly, this hasn't changed much over the decades. Back in the 1930s, once you reached 65 you'd have a life expectancy of only three years less than the contemporary figure.[4] So modern medicine has only added three more years of life! According to the American Academy of Anti-Aging Medicine, this is still a long way short of our true potential as human beings, which they believe is living to the maximum lifespan of 120, free of degenerative disease.

As Figure 1 (below) shows, the ideal health objective is to flatten the lifespan curve, so you live fully, with all your senses intact, free of pain and disabling fatigue, until you die. For most of us, realistically, that point should be around 100 years old, at least. This is what is predicted for those who have followed the practice of 'optimum nutrition' and a healthy lifestyle all their lives. If you've only started behaving yourself in your forties or fifties, your maximum lifespan may be in the nineties. But we are not talking about adding years of decrepitude. We are talking about having the energy, memory and physical function of a 50-year-old when you are 70, and of a 60-year-old when you're 90.

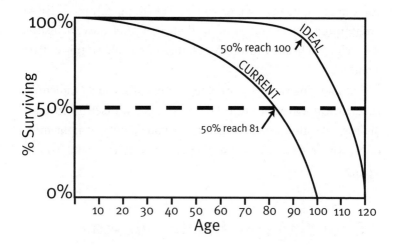

Figure 1. Extending your healthy lifespan

How can this be done? One of the best ways to extend your healthy lifespan is by reducing your homocysteine level with the H Factor programme. With every five-point decrease in your H score, you will gain:

- A 49% reduced risk of death from all causes

- A 50% reduced risk of cardiovascular death (such as heart attacks and sudden death)

- A 26% reduced risk of death from cancer

- A 104% reduced risk of death from any causes other than cancer or heart disease.

These are the extraordinary findings from a comprehensive research study at the University of Bergen in Norway, published in 2001 in the *American Journal of Clinical Nutrition*.[5] They measured the homocysteine levels of 4,766 men and women aged 65 to 67 back in 1992, and then recorded any deaths over the next five years, during which 162 men and 97 women died. They then looked at the risk of death in relation to their homocysteine levels. Remarkably, they not only reconfirmed the relationship between heart attacks, strokes and high levels of homocysteine, but also found that 'a strong relation was found between homocysteine and all causes of mortality'. In other words, your level of homocysteine is an accurate predictor of how long you are going to live, whatever the eventual cause of death may be!

If you are already in your fifties or sixties, you might be tempted to view this news with gloom. But with a guaranteed way out – the H Factor programme – the news is all good, because you can begin to do something about it right now.

Now, let's take a look at more encouraging news that shows how homocysteine predicts each of the top five killer diseases. Why is this encouraging? Because it means you can dramatically reduce your risk of ever dying from any one of these by lowering your homocysteine level.

Heart attacks – reduce your risk by 80%

Elevated homocysteine causes at least 56,000 deaths in the US each year. Other medical experts now strongly feel that 12 per cent of all deaths can be attributed to high homocysteine; in the US that would amount to 240,000 deaths a year. By some estimates, 30 to 40 per cent or more of all heart attacks or strokes may be the result of elevated homocysteine. That's the bad news.

The good news is that by following our plan, you can reduce your risk by up to 80 per cent.

Strokes – cut your risk by 82%

Strokes are the third biggest killer after heart attacks and cancer, and often leave survivors seriously disabled. Yet the fact is that your homocysteine level predicts your risk of a stroke even better than it predicts your risk of a heart attack – and it does so better than any other single measure, including your cholesterol level, blood pressure or whether or not you smoke. The upside is that knowing and reducing your H score can cut your risk by up to 82 per cent.

Cancer – cut your risk by a third

Cancer is largely preventable. The World Cancer Research Fund estimates that by eating the right diet you can cut your risk of cancer by up to 40 per cent. The European Commission estimates that a quarter of a million lives could be saved each year across the 12 member states through dietary changes alone. You can further cut your risk by eating organic whenever possible, exercising regularly, adding vitamin and mineral supplements and avoiding known carcinogens – such as smoking, oestrogen replacement therapy (HRT), and eating excessive animal fats and overcooked or blackened meats.

We estimate that cancer is about 85 per cent preventable. Research published in the *New England Journal of Medicine* describing a study involving 45,000 pairs of twins found that cancer is more likely to be caused by diet and lifestyle choices than by genes. Identical twins, who are genetically the same, had no more than a 15 per cent chance of developing the same cancer. This suggests that the cause of most cancers is about 85 per cent environmental – that is, down to factors such as diet, lifestyle and exposure to toxic chemicals. This study found that choices about diet, smoking and exercise accounted for 58 to 82 per cent of cancers studied.[6]

So where does homocysteine come into all this? Cancer is triggered in large part by damage to DNA – and having a high homocysteine level means your DNA is more vulnerable to damage, and poorly repaired once damaged. At the other end of the scale, a high level of homocysteine has been found to be a very good indicator of whether cancer therapies are working. The homocysteine level rises when tumours grow, and falls when they shrink. Forms of cancer already clearly linked to high homocysteine include cancer of the breast and colon, and leukaemia, among others. Low homocysteine is likely to reduce your risk of these by a third. Coupled with other diet and supplement changes, you should be able to cut your cancer risk by more than half.

Diabetes – lower your risk

Type II, or adult-onset, diabetes is highly preventable. Yet more and more young people are developing it. The obesity 'epidemic' in the West has helped fuel this rise. If you are obese, the risk of developing diabetes goes up 77 times! Why? Because too little exercise, not enough fresh fruits and vegetables, too many refined carbohydrates (white bread, cakes, biscuits, sugar, etc.) and the added stress on your body all make your blood sugar control go out of balance.

When blood sugar levels rise too high, the body produces insulin to help bring the level down. The more often you have these high levels of glucose (that is, high blood sugar), followed by insulin production, the more 'deaf' or resistant your body cells become to insulin. Insulin resistance is now found in a quarter of all people in the industrialised world, and in over 90 per cent of obese people. Since insulin lowers blood sugar, the net result of insulin resistance is too much sugar in the blood – in other words, diabetes.

Diabetics are at risk of having high homocysteine because we now know that the abnormally raised insulin seen in most diabetics stops the body from lowering and maintaining a healthier homocysteine level. By following the H Factor programme, you will be able to help to reduce your risk of diabetes or, if you are diabetic, you'll be able to help keep it under better control and reduce complications.

Alzheimer's – halve your risk

The evidence indicates that if you can lower your H score, you will significantly lower your risk of getting Alzheimer's. Homocysteine is strongly linked to damage in the brain. Dr Matsu Toshifumi and colleagues at Tohoku University, Japan, conducted brain scans on 153 elderly people and checked them against each individual's homocysteine level. The evidence was crystal clear – the higher the homocysteine, the greater the damage to the brain.[7]

A recent study in the *New England Journal of Medicine* charted the health of 1,092 elderly people without dementia, measuring their homocysteine levels. Within the next eight years, 111 were diagnosed with dementia. Eighty-three of this group were diagnosed with Alzheimer's. Those with high blood homocysteine levels (in this study, above 14 units) had nearly double the risk of Alzheimer's. All this strongly suggests that following the programme outlined in Parts 4 and 5 of this book should, at the very least, halve your risk of developing Alzheimer's in later years.[8] The chances are excellent that the H Factor programme will eliminate your risk completely.

****

And the upshot of all this? The connection between homocysteine and the 'big five' killer diseases indicates that if you can
lower and maintain your H score into the superhealthy range,
defined as under 6 units – and you can with the H Factor programme – you are likely to add at least ten or more quality years
to your life.

MEAT CHEESE + SOME OTHER PROTEINS RICH IN

2

VITAL AMINO ACID

What is homocysteine?

MADE IN BODY FROM ANOTHER AMINO ACID. METHIONINE

WE NOW KNOW that homocysteine is an extremely important marker for health and disease. But there's much more to tell. This chapter is a full-length portrait, if you like, of this vital amino acid.

■ Methyl magic

Homocysteine is made in the body from another amino acid, methionine. As meat, cheese and some other proteins are especially rich in methionine, we tend to eat this amino acid every day.

Why does the body make homocysteine and what does a high level tell us? It's all to do with a fundamental process upon which your life depends. We first encountered it in the introduction to this book. It's called methylation.

To understand methylation we need to know a bit about body chemistry. You eat 10 tons of food in your lifetime and, somehow,

[handwritten annotations in top margin: "Body – Concocted of Millions of Chemicals / Glucose to Fats / Amino Acids to Hormones + Neurotransmitters"]

this turns into you. Your body is quite literally a sea of chemicals, a hairy bag of salty soup concocted out of millions of them, from glucose to fats, and amino acids to hormones and neurotransmitters.

For example, when you are under stress, the body makes more adrenalin to keep you going. When you go to bed, the body releases melatonin to help you sleep. When you've got a cold or flu the body makes more glutathione, which turns your immune cells into cold-busting warriors. These are just three examples of literally hundreds of thousands of adjustments the body makes every second to keep you healthy and alive.

But how on earth does the body keep everything in balance? This is where methylation comes in. In the methylation process 'methyl groups', which are made of one carbon and three hydrogen atoms, are added to, or subtracted from, other molecules. This is how the body actually makes the substances it needs, or breaks down those it doesn't – by transforming one biochemical into another.

Methylation happens over a billion times a second. It is like one big dance, with biochemicals passing methyl groups from one partner to another.

Take noradrenalin. The brain produces this chemical to keep you happy and motivated. However, if you are under stress, it adds a methyl group to noradrenalin in the adrenal glands to make adrenalin, which gives you a burst of energy and aggression known as the 'fight or flight syndrome'.

This is how homocysteine is made in the body. When you eat a piece of fish containing methionine, it's incorporated into your bloodstream and inside your cells and a methyl group is taken away from the methionine, leaving you with homocysteine. Ideally, the body adds a different methyl group back to homocysteine to convert it into an extraordinarily important chemical called S-adenosyl methionine (SAMe, pronounced 'Sammy', for short). SAMe is a natural anti-depressant, anti-arthritic and liver-protecting agent in your body. It also becomes a methyl donor in its own right, readily giving up its methyl group to help alter other body chemicals.

Homocysteine can also be converted to another extremely important body chemical, glutathione. Glutathione is the body's best anti-ageing antioxidant and detoxifying agent. A low glutathione level is, like a high homocysteine level, linked to increased risk of death from all common causes. So methylation is also the key to slowing the ageing process and keeping your body free of toxic chemicals.

It is also thought that methylation plays a critical role in protecting us from certain serious diseases. Methyl groups are added to and subtracted from our DNA. When not enough methylation is going on, our DNA cannot properly repair itself, which puts us at higher risk from cancer and autoimmune diseases such as rheumatoid arthritis or lupus.

■ What's your methyl IQ?

It can be helpful to think of people with high-functioning methylation as having a high methyl IQ. They stay in balance, while those with a low methyl IQ suffer from chemical imbalances that ricochet into almost every organ and tissue of the body.

The best, most sensitive methyl IQ test is your homocysteine level. When your H score is low (below 6 units), you are well methylated, your SAMe and glutathione levels are most likely high, and you are in good health. When your H score is too high, you suffer from a methyl deficiency, and not surprisingly a deficiency in SAMe, glutathione and lots of other vital biochemicals.

This relationship between homocysteine, methylation and vital body chemicals is complex but vital (see Figure 2 below). Provided your body has a good methyl IQ, only small amounts of homocysteine accumulate, with the great majority immediately methylated, en route for greater destinies.

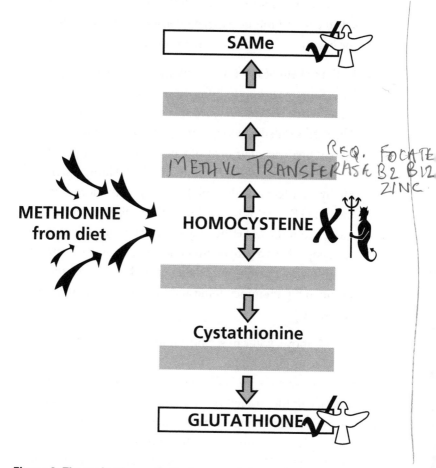

Figure 2. The main players in the homocysteine pathway

(For the more technically minded, turn to Appendix 1, page 236, for an unabridged version of the homocysteine pathway.)

We've had a taster on how the body deals with homocysteine. Now let's look closer. Normally, as soon as methionine is converted to homocysteine, the body then changes it as quickly as possible in one of two ways:[9]

1 It adds sulphur with an enzyme called cystathionine beta-synthase, which turns homocysteine into something called cystathionine. From there, with the help of another enzyme called cystanthionine lyase, it's converted into glutathione, the benefits of which we've seen above.

2 With the help of key B vitamins and zinc, the brain adds back a methyl group to homocysteine in a process called remethylation. Two enzymes are essential to this process. The first is called homocysteine methyltransferase and the second methylenetetrahydrofolate reductase – impossible to remember, so it's abbreviated to MTHFR. Homocysteine methyltransferase adds a methyl group to homocysteine, then MTHFR helps turn it into SAMe. (You'll be hearing more about the MTHFR enzyme later because about one in ten people have a genetic mutation that means this enzyme doesn't work so well, and so these people are much more prone to high homocysteine and have to work a bit harder to reduce it.)

In Figure 2 above, you can see how homocysteine can convert to either SAMe or glutathione.

If this conversion process isn't working well – for example, due to a lack of the co-factors (vitamins and nutrients) which the homocysteine-converting enzymes need to function – homocysteine begins to accumulate in the body, and that spells trouble. Increased levels of homocysteine and therefore decreased levels of methylation, SAMe, glutathione and B vitamins are associated with chronic symptoms many of us experience every day. We'll be looking at these in Chapter 6, which will also tell you how to assess your level of homocysteine.

What if you feel good, though? Don't make the fatal mistake of assuming that if you don't have the symptoms, you don't have a problem with homocysteine. High homocysteine, especially in the early stages, is often symptom-free, exactly like many of the serious medical conditions associated with it, such as heart attacks, high blood pressure and strokes.

In any case, lowering your H score will be a priority. And intelligent nutrients are one way of doing it.

■ Intelligent nutrients

The reason homocysteine accumulates in the body is because enzymes, the chemical catalysts in the biochemical transformation process, aren't working properly. Have a look at Figure 3. Here you can see the spotlight on the enzymes that keep your brain, liver and other body organs doing the right thing with sulphur and methyl groups.

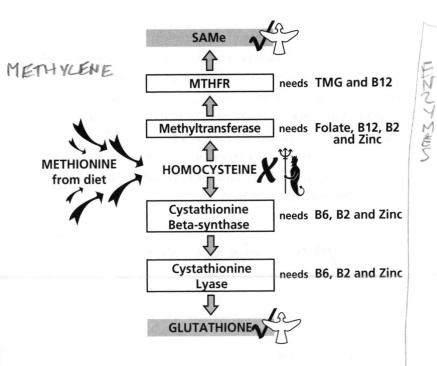

Figure 3. The main enzymes in the homocysteine pathway

Look at the diagram and you'll see that these enzymes don't work alone. They have helpers, called co-factors: primarily the vitamins folic acid (folate), pyridoxine (B6), cobalamin (B12), riboflavin (B2) and the mineral zinc. Among them is also a very special nutrient called TMG (for trimethylglycine), of which more later.

These are listed below:

Enzymes and their co-factor nutrients

Enzymes	Co-factor nutrients
Cystathionine beta-synthase	B6, B2 and zinc
Cystathionine lyase	B6 and zinc
Homocysteine methyltransferase	B12, folate and TMG
MTHFR	B12, B2, folate and zinc

So as you can see, the dance of homocysteine is mainly choreographed by these co-factors. The B vitamins among these are 'intelligent' in that they help you stay chemically flexible, well methylated, and in the best possible physical and mental health. As you will discover in Part 5, very few of us get enough of these nutrients from our diet, and certainly not enough to lower a high homocysteine level to an optimal level. The average intake of zinc, for example, is 7.5mg, which is only half the basic recommended daily allowance, or RDA.

■ Eat right *and* take your vitamins – and more

In the US, where homocysteine is now seen as superseding cholesterol as the best predictor for a heart attack, more and more health consumers are becoming B-vitamin crazy. And quite rightly, too. The American Medical Association published a report in 2001 suggesting that if every cardiovascular patient were to supplement B12 and folate, no less than 310,000 lives in the US alone would be saved in the next ten years. Meanwhile, the *New England Journal of Medicine* ran a lead editorial written by top cardiologists entitled 'Eat Right and Take a Multivitamin', again arguing that optimal intakes of these vital B vitamins can dramatically cut heart disease risk.

Some critics continue to disparage daily vitamin and nutrient supplements, although this is rapidly becoming an old-fashioned view among the medical community. As you will see, however, the evidence shows clearly that eating a well-balanced diet is not enough in itself to lower high homocysteine to a safe level.

But simply adding ordinary multivitamins to the equation won't do the trick, either. In fact, it could literally be a fatal mistake. Why? Because the evidence shows, again and again, that if you have a high homocysteine level you'll need more than basic amounts of B vitamins. You'll also need 'methyl donors' – nutrients like TMG and SAMe (both of which are available as supplements), which can donate an abundance of methyl groups to your body's chemical dance. And you need to remember to test your H levels regularly.

The story of a 60-year-old man treated by the Life Extension Foundation in the US is a case in point. This man had bypass surgery and was suffering from chest pains caused by a reclogging of his coronary arteries, a very common occurrence in heart surgery patients with high homocysteine. He was well aware of the dangers of high homocysteine and was supplementing 100mcg of folate a day (that's almost 1,000 times the RDA!) plus other key homocysteine-lowering B vitamins. He was smart enough to have his H score retested. When he did, he discovered that his homocysteine level, although lower, was still in the extremely high-risk range, above 15 units. So, in addition to the key B vitamins, he began to take 6g of TMG. His H score then dropped dramatically to 4, indicating zero risk.

This story shows why blindly supplementing with the big four B vitamins, and failing routinely to retest your homocysteine level or reassess your diet, can literally be a fatal mistake. If you want to live long and stay healthy, you need to know whether your regime is working for you, and what to do if it isn't. We will tell you how, step by step, later on in Parts 4 and 5, 'The H Factor Diet and Lifestyle' and 'The H Factor Supplements'.

■ Are you well methylated?

So, in addition to the vital big four 'methyl movers' – that's folate, B6, B12 and B2, plus zinc – we all need an abundance of the methyl groups themselves. These, as we've seen, are dispensed by methyl donors such as SAMe and TMG.

SAMe is not necessarily the best nutrient to supplement, although it is very important within cells. As a supplement it is, among other things, both very unstable and very expensive to produce. In addition, lowering homocysteine naturally increases SAMe levels. (Its special qualities are discussed in more detail on page 212 of Part 5.) Instead, TMG is the single best and most affordable methyl donor discovered so far. And as you will see, in combination with the big four B vitamins, it's the best homocysteine buster. TMG, as well as being available as a supplement, is formed naturally from choline, which is found in fish and eggs and so is easy to get from our diets – assuming we're not strict vegetarians! We'll be talking a lot about these nutrients later in the book.

To summarise, a good starting point is to take the big four B vitamins and zinc, and make sure you are getting enough TMG and SAMe.

Beyond homocysteine

These nutrients have amazing effects not only on your homocysteine levels but also on your day-to-day health. When levels are optimised, you'll find you have:

- Improved mood, memory and mental clarity

- Improved liver function

- Better condition of hair, skin and nails

- Increased energy

- Better sleep

- Reduction in arthritic pain

- Raised glutathione level (slowing the ageing process and aiding in detoxification)

- Dramatic reduction in risk of heart disease, strokes, cancer, and Alzheimer's.

And, of course, these nutrients can also dramatically lower your homocysteine risk within weeks, as part of the H Factor programme. So, you'll feel better AND dramatically reduce your risk of disease.

We've looked in depth at homocysteine's place in body biochemistry. But to fully understand how vital it is to regulate this amino acid, let's have a detailed look at what high levels of it can do to the body.

3

Ten reasons to lower your homocysteine

IN THE LAST CHAPTER we saw that the lower your H score, the better your body is at keeping the perfect balance of biochemicals, making you 'well-tuned'. This means more energy, more stamina and endurance, clearer thinking, fewer infections, and better skin and weight control.

To give you a sense of how homocysteine, as a marker for your methyl IQ, is central to how your body and brain keep themselves healthy, here are ten good reasons for keeping it low and in balance.

■ 1. Speeded-up oxidation and ageing

In the 1980s the spotlight fell on antioxidants such as vitamins A, C and E and the minerals zinc and selenium, all of which were found to have an important role in preventing ageing. When you process just about anything in your body, from food and drink to the air you breathe, you end up with damaged forms of oxygen called oxidants or 'free radicals'. The body makes a bucketful of oxidants every day just from processing oxygen and breaking down glucose, the body's main fuel. If you smoke, eat fried, browned or burnt food, exercise excessively, breathe in exhaust fumes, are exposed to strong sunlight, even suffer from chronic inflammation (which creates a vicious cycle of oxidant damage), you'll end up with even more. These oxidants damage your skin, your lungs, your digestive tract, your brain, your arteries and your DNA, the genetic code that keeps cells behaving properly. Oxidation triggers many diseases, including heart disease, strokes, cancers, and autoimmune disease such as rheumatoid arthritis[10] and diabetes.

We now know that high homocysteine dramatically and directly increases oxidation, and the damage caused by it.

▪ 2. Damage to your arteries

Increased oxidation is one way, but not the only way, that homocysteine damages your arteries. High homocysteine blood levels can cause blood cholesterol to change to a very dangerous form, called oxidised LDL cholesterol, which can severely attack artery walls.

Once there's damage to your arteries, other types of cells begin to stick to your arterial walls. Among these are macrophages – immune cells that normally help to repair damage – and these are thought to be involved in developing atherosclerosis or the thickening of artery walls. In the presence of high homocysteine, macrophages become a lot stickier, causing ever thicker deposits in the arteries.[11] When this happens, all it would take for you to have a heart attack or stroke would be for a big blood clot to stick. And guess what? Homocysteine makes your blood clot much more easily than it should, increasing the risk of dangerous clots and sudden heart failure.

Raised homocysteine also affects the flexibility of arteries, flexibility being one of the ways the body copes with any arterial clogging. It does this indirectly in two ways. When oxidants, pushed by high homocysteine levels, damage the lining of arteries, the arteries begin to stiffen and the loss of flexibility causes blood pressure to rise dangerously high. High homocysteine (and low levels of folic acid) also affects a gas that's crucial to arterial flexibility: nitric oxide (NO). Some modern heart and blood pressure drugs are based on boosting NO activity in arteries.[12] Homocysteine is now known to profoundly lower NO levels.[13] When combined with increased oxidation, this makes a person with a high H score a prime candidate for a stroke or heart attack.[14]

In fact, having a high homocysteine level increases your risk of a heart attack or stroke by as much as 70 per cent. Depending on where one sets the safe range for homocysteine levels, at least 40 per cent of people with a history of strokes have abnormally high homocysteine levels, while at least 30 per cent of people with a history of heart attacks have very high homocysteine levels. Of those 'at risk' of developing these diseases – people with high blood pressure or a family history of strokes or heart attacks, for example – at least 20 to 30 per cent have high homocysteine levels. That's a lot of people. We are conservatively talking about 1 million people in Britain and up to 8 million in the US.

Despite all the evidence we've outlined here, virtually no doctors and far too few cardiologists routinely test for homocysteine at the moment. However, we confidently predict that within the next ten years or so they all will. But with the abundance of published scientific evidence and ease and safety of treatment, who wants to wait that long?

■ 3. A weakened immune system

Whether you are fighting a virus like the common cold, a bacterial infection (perhaps picked up from something you ate on holiday), an allergy or a rogue cancer cell, you need an immune system that's fighting fit.

Your immune system attacks these invaders, but it is in turn attacked by oxidants. As we saw above, these are everywhere and can do extreme damage if there are high levels of them. So the more antioxidants your immune army has available, the stronger it becomes and the more able to stop the damage wreaked by these invaders.

The most important and powerful antioxidant of all is glutathione, which we met in Chapter 2. Glutathione is an essential sulphur-containing compound that lives inside your immune cells – in fact every cell – acting much like a benevolent police force. Oxidants also get produced when cells break down glucose to make energy, and glutathione helps mop up the cell's own 'exhaust fumes'.

Since, as we've seen, excess oxidants speed up ageing, having high levels of glutathione inside your cells is some of the best health insurance you can have. Remember – lowered glutathione, like elevated homocysteine, is associated with an increased risk of early death from all common causes.

So how do you boost your glutathione? One excellent way is to lower your homocysteine to a safe range. As you can see in Figure 2 on page 23, if you have a high homocysteine level, you are not making enough glutathione. With the right intake of B vitamins, zinc and methyl donors like TMG and SAMe, from both diet and supplements, you can raise glutathione levels.

■ 4. Damage to your brain and a lowered IQ

One of the most sensitive organs of the body is your brain. Because it receives so much of your body's available oxygen from the blood supply and is active even when you are asleep or at rest, it is highly prone to oxidation – even more than your arteries or your joints when inflamed and damaged. For example, Alzheimer's is essentially an inflammatory disease, indicating increased oxidation. Even clinical depression and attention deficit hyperactivity disorder (ADHD) are linked to excessive oxidation.

The brain's communication chemicals are what keep us alert, happy, calm and connected, and these all depend on methyl groups and proper methylation, which we explored in Chapter 2. Methylation keeps the brain's chemical messengers, the neuro-transmitters, in balance, and so too brain function. But as we have learned, the 'dance' of methylation depends on getting enough B vitamins, zinc and essential nutrients such as SAMe and TMG. So supplementing these in the right combination, which we'll learn about in Parts 4 and 5, will keep the workings of your brain in top condition.

What about IQ, though? Back in the 1980s, we extensively tested the effects of raising schoolchildren's IQs using multivita-mins. We proved, in double-blind trials published in the *Lancet*, that the simple addition of optimal amounts of vitamins and minerals, and especially B vitamins, could boost the average child's IQ score by nine points![15] We later found that those chil-dren with the lowest intake of B vitamins before being supple-mented had the greatest improvement. And isn't it fascinating that among the B vitamins most effective in raising a child's IQ, it's the homocysteine-lowering folate, B2, B6 and B12 that stand out. If only we'd known more about homocysteine back then! Perhaps we would have found that the lower a child's homo-cysteine, the higher their IQ. That research has yet to be done.

■ 5. Increased pain, inflammation and blood clotting

When the body is being insulted in some way, it reacts with redness, pain, tenderness, heat and swelling. This so-called inflammatory reaction occurs when you sprain your ankle or get stung by a bee, for instance. People suffering from arthritis or migraines also experience an inflammatory reaction. Most painkillers are anti-inflammatory drugs, helping to turn off the 'red alert' signal of the body. The main inflammatory messenger of the body is a hormone called prostaglandin E2, or PGE2, which is derived from an animal fat called arachidonic acid.

High levels of arachidonic acid and PGE2 not only indicate a state of inflammation or emergency. They are also strongly linked to all sorts of disease, from heart attacks to strokes, Alzheimer's, cancer, and autoimmune diseases such as diabetes and arthritis. While inflammation is the body's short-term way of coping with an insult – for example, by isolating the poison from an insect bite by swelling, or stopping you from walking on that twisted ankle – prolonged inflammation actually damages joints, arteries and nerves.

High homocysteine levels promote arachidonic acid release, an increase in PGE2 and hence inflammation.[16] In the next part of this book, you'll see how high homocysteine levels are linked to most inflammatory diseases.

■ 6. Vulnerability to cancer and problems with detoxification

Glutathione is essential in detoxifying the body and helping repair damaged DNA. So if you have a high H score, and therefore low glutathione, you will increase your risk of getting many kinds of cancer, and of premature cell death. Just about every cancer known is linked to glutathione deficiency, and many are linked to high homocysteine. Glutathione, along with SAMe, is your liver's best friend, too. Any insult to the body – smoking, drinking, allergens, viruses, chemicals or drugs (both illegal and prescribed) – increases your need for glutathione and dramatically raises your homocysteine level.

You may have a glutathione deficiency if you have smoker's cough, chronic bronchitis, asthmatic coughing and wheezing, difficulty concentrating, frequent headaches, food allergies and cravings, joint pain, muscle pain, frequent tiredness, irritability and mood swings, or recurrent colds and other infections.

▪ 7. A faster-ageing brain

In most people, 20 per cent of the 100 billion or more brain cells die over a lifetime. When you reach 70, your brain will have shrunk by 10 per cent. With this shrinkage often comes a gradual loss of control of the complex orchestra of hormones and neuro-transmitters that keep you on the ball. The result can be diminished brainpower, slower memory retrieval, reduced sex drive, poor sleep, less energy, less motivation, chronic depression, social isolation and fewer highs.

After the age of 75, one in ten people develop dementia, the most common form of which is Alzheimer's disease, characterised by a rapid loss of brain cells. This happens because DNA within brain cells gets oxidised and damaged by something called beta-amyloid.

As you'll see in Chapter 13, the risk of developing Alzheimer's or just a declining memory is very strongly linked with high homocysteine. The big question is why. Recent research by Dr I. I. Kruman and colleagues from the Gerontology Research Center at the National Institute on Aging in the US may have the answer.[17] Not surprisingly, they found that the higher the homocysteine, the less methylation and the less folate in the brain. The brain repairs its DNA via methylation, so low folate with resulting high homocysteine and poor methylation means more beta-amyloid-damaged brain cells and less effective repair of this damage. A double whammy.

Happily, you can protect and rejuvenate existing brain cells at any age[18] provided you have a safe level of homocysteine, and the right nutrients. This, along with the right lifestyle and attitude, means that age-related memory loss needn't happen to you. Research clearly shows that many healthy elderly people show no decline in mental function right up to death, and no increased rate of brain shrinkage even after 65.

▪ 8. Hormonal problems

Although we do not yet know exactly why, high homocysteine is strongly linked to oestrogen deficiency, which is particularly prevalent in post-menopausal women. This may also explain why the risk of heart disease is so much greater in women once they've passed the menopause. The fall-off in oestrogen leads to an increase in homocysteine and thus an increase in heart attack and stroke risk.[19]

Conversely, raising oestrogen levels in post-menopausal women tends to lower homocysteine. A recent study by Dr Hong Tao from the Department of Cardiology at Peking University First Hospital gave both men and women with coronary artery disease low doses of oestrogen. After six weeks they had, on average, a highly significant 14 per cent decrease in their homocysteine levels.[20] There was also a highly significant decrease in oxidised, or damaged, LDL (bad) cholesterol. So having enough oestrogen is good news as far as homocysteine is concerned. However, it is premature to recommend oestrogen HRT as a means to lower homocysteine, especially in the light of the known increased risk of breast cancer associated with HRT (see Chapter 30, 'Correct Oestrogen Deficiency').

There's quite a lot of accumulating evidence that lowering homocysteine may also help to keep your hormones in balance. In other words, it works both ways. This is discussed in Chapter 17, where you'll also find out that miscarriages and problem pregnancies are strongly linked to high homocysteine levels.

■ 9. Deficiency in B vitamins

It is now proven beyond a doubt that, as we've seen, a lack of certain B vitamins raises homocysteine, and that enough B vitamins, plus zinc and methyl donors, can lower homocysteine into the superhealthy range. What this means is that your homocysteine level is now considered by many authorities to be the single best, most sensitive indicator of your B12 and folate status, and perhaps of your B6 status as well. Rather than shooting in the dark and assuming that your 'well-balanced diet', plus that daily multivitamin, are giving you all the B vitamins you need, you can find out for a fact by testing your homocysteine level.

■ 10. Deficiency in SAMe

As well as being the body's main methyl donor, SAMe is the brain's master tuner. It helps keep you happy. In fact, there are over 100 double-blind studies in which volunteers given SAMe or a dummy pill responded as well to SAMe as to anti-depressant drugs, or even better.[21] According to one review of these studies, 92 per cent of depressed patients responded to SAMe and experienced far fewer side-effects, compared to 85 per cent taking standard anti-depressant medications. SAMe also works fast. 'Most people taking SAMe see some effect within 10 days,' says Dr Teodoro Bottiglieri, director of neuropharmacology at Baylor University Medical Center in Dallas, Texas.

Of course, since SAMe is the brain and body cells' most important methyl donor, you'd expect a wide range of benefits. And that's exactly what you'll find. SAMe helps restore joint mobility and relieves pain in arthritis,[22] helps reduce the muscle pain of fibromyalgia,[23] helps enhance liver function[24] and protect against liver disease, and helps protect the brain from damage caused by oxidation.[25] Roll on SAMe!

How do you know if your body is making enough SAMe, and how do you encourage it to make more? Like glutathione and B vitamin status, your homocysteine level is also one of the very best indicators of your SAMe status. The lower the level, the more SAMe your body makes. With enough methyl groups, B vitamins and zinc around, up goes your SAMe (and glutathione) and down goes your homocysteine.

The moral of this story is simple. If you want to be in optimal health, with the longest, healthiest life possible, get your homocysteine level into the superhealthy range by following the H Factor programme.

PART 2

How's Your Homocysteine?

4

Homocysteine – your most vital statistic

THERE'S AN OLD SAYING about the acceptance of new ideas. First sceptics and critics say, 'It's not true and not important.' Then they say 'It's true, but not important.' Then they say 'It's true, it's important, but it's not new!' This describes exactly what has happened with homocysteine. As we saw in Part 1, Dr Kilmer McCully struggled on the sidelines for over 20 years, waiting for the medical profession to take his discovery seriously.

Then, in the mid-1990s things started to heat up. There was the Massachusetts-based Framingham Heart Study in 1995, which found that having more than 11.4 units of homocysteine in the blood dramatically increased the risk of heart disease.[1] Then, in 1997, a study published in the *Journal of the American Medical Association* proved that homocysteine was more important than cholesterol in predicting heart attacks and at least one in five people with a history of heart disease had 'high' levels. Homocysteine had entered the 'true, but not important' phase.

In November 2002, a 'meta-analysis' of 92 of the best published scientific studies on homocysteine confirmed that homocysteine is not only a major risk factor for heart disease, but that it actually causes heart disease.[2]

We believe that homocysteine is not only important, but that it is the *single most vital health statistic*. More accurately than any other single measure, it can predict where you are on the scale of health, and whether you have a significant risk of all the major killer diseases, plus many, many others. It is more important than your weight, your blood pressure, your blood type or your cholesterol level. We predict that, within five to ten years, homocysteine will be a routine part of your medical check-up, right alongside cholesterol, blood pressure and blood sugar, and will be tested for by *all* knowledgeable and conscientious doctors in anyone suffering from or suspected of being at risk of disease.

The good news is that you don't have to wait five years. You can take a laboratory test to find out your H score now – either through an enlightened doctor, a nutrition consultant or even by yourself with an easy-to-use home test kit. You'll find the details in Chapter 7.

As we outlined in Part 1, we also believe it is a potentially fatal mistake to assume that your homocysteine level is OK without testing just because you are symptom-free, eat a well-balanced diet or take vitamin supplements that include B vitamins. Some people, for both lifestyle and genetic reasons, are much more prone to high homocysteine and more resistant to conventional therapies than others. These risk factors are explained in detail in Chapter 6.

In a nutshell, our review of over a thousand scientific research papers on homocysteine makes it clear that we all need to know our homocysteine level: it is of crucial importance in helping you identify what you personally need to do to stay in optimal health and free from disease.

■ How much is too much?

So, what level of homocysteine is healthy and what levels indicate what kind of risk? As more and more evidence is published in medical journals, the 'healthy' homocysteine level is moving lower and lower.

Below 15?

For many years a homocysteine value below 15 units was considered to be safe. This level has since been discarded almost universally among homocysteine researchers. It's much too high, and therefore much too dangerous.

Above 14 units, you not only have a major risk of heart disease and strokes, but also double the chances of getting Alzheimer's later in life. Above 10.4 units, risk of certain cancers starts to increase.

The Framingham Heart Study, remember, found that over 11.4 units of homocysteine in the blood significantly increased risk of heart disease.[3] Many authorities have adopted this level as safe when evaluating patients. For example, here's what the American Academy of Family Physicians says:

> A healthy homocysteine level is less than 12 units. A level greater than 12 is considered high. If your homocysteine level is 12 to 15 units and you have blockages in any blood vessel, you need to lower your homocysteine to less than 12. If you have no other major risk factors for cardiovascular disease and you do not have atherosclerosis, it may be okay for you to have a modestly high level of homocysteine.

We emphatically disagree with this statement and think it is dangerous for reasons that will become clear.

Other leading cardiology researchers think the optimal level is likely to be lower than the Framingham finding. A case in point is Dr Johan Ubbink and his team from South Africa's University of Pretoria. They compared those with risk of cardiovascular disease and those not at risk and determined the optimal homocysteine levels to be somewhere between 4.9 and 11.7 units. This is the range found in various studies to have the lowest risk and incidence of heart attacks, strokes and peripheral artery disease.

Below 9?

It has been found that cardiovascular patients whose H score is lower than 9 units and who have needed to have coronary angioplasty to unblock their arteries have significantly fewer new arterial blockages – which are a disturbingly common problem following heart surgery – and fewer cardiovascular disease complications later.

This is consistent with what research shows about mothers and babies. Mothers who give birth to healthy, normal babies consistently have homocysteine values lower than 9 units. Above 9 units there is an increased risk of babies being born with congenital defects like cleft palates, club feet and spina bifida, or problems in pregnancy, including recurring miscarriages.

The strongest evidence for setting a 'safe' homocysteine level at below 9 units comes from Norway. Researchers at the University of Bergen, the Department of Heart Disease at Haukeland University Hospital, and the National Health Screening Service in Oslo – widely recognised as one of the leading homocysteine research groups in the world – looked at the chances of 55-year-olds with coronary artery disease surviving five years, based solely on their homocysteine levels.[4]

As the graph below shows, the Norwegian study showed that those with homocysteine levels below 9 had a 95 per cent chance of surviving five years or longer. Those with a level between 9 and 14.9 had approximately an 80 per cent chance of surviving five

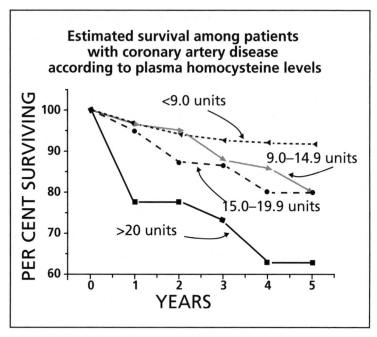

Source: Adapted from a University of Bergen study

Figure 4. Safe homocysteine levels are less than 9 units

years, while those with a homocysteine score of 15 or more had a 75 per cent chance of surviving. Above 20 units the chance of surviving five years drops to 60 per cent.

Below 6?

Even a homocysteine score of 9 may be too high. According to a large-scale survey published in the American Heart Association's journal *Circulation*,[5] any homocysteine score above 6.3 means a greater risk of cardiovascular disease. This survey looked at the homocysteine levels of several thousand people and compared those with their risk of cardiovascular disease. As the graph below shows, there is no safe so-called 'normal range' for homocysteine; that is, what is 'normal' for most people can be unhealthy for others and places them at risk. However, when

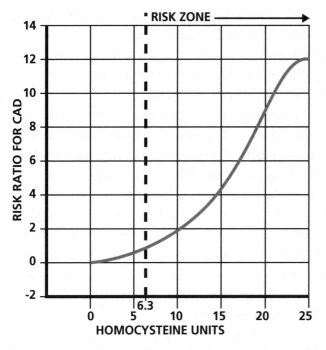

(CAD = coronary artery disease; 0 = zero risk; 2 = double the risk)

Source: Adapted from K. Robinson et al., *Journal of Circulation*, vol. 92 (1995), pp. 2825–30

Figure 5. The relative risk of heart disease compared to homocysteine

homocysteine is maintained below 6.3 units, the risk of coronary artery disease is virtually zero.

Another important, extremely revealing, study found that for every 3-unit increase in homocysteine, your risk of a heart attack increases by 35 per cent.[6]

Health, however, isn't simply an absence of disease. It's also about wellbeing and good internal biochemistry. If a homocysteine level below 6.3 indicates minimal risk of a heart attack, it is highly likely that the superhealthy level is below this. Also, there's currently lots of information about homocysteine levels and the risk of heart disease, strokes and peripheral artery disease, but very little being reported to us about the other 100 diseases now linked to homocysteine. What level, for example, is related to the lowest risk of cancer, chronic fatigue, diabetic complications, arthritis, problem pregnancies or memory problems? The truth is, we should know, but we don't.

One of the first research groups to take homocysteine seriously was the US-based Life Extension Foundation, which we first read about in Part 1. For more than a decade they have been advising their members to take homocysteine very seriously, and have recommended supplementing optimal amounts of B vitamins, especially B6, B12 and folate, in the belief that these alone would ensure a low, safe level of homocysteine. However, over the last decade they have identified several cases where people were suffering from coronary artery disease and had lethal levels of homocysteine, despite taking their recommended doses (and higher) of vitamin supplements.

In response to these particular cases, the Foundation analysed all the homocysteine tests it had conducted on its members and discovered that 62 per cent of them tested still had too much homocysteine in their blood, even when supplementing with the recommended B vitamins. The following chart shows the breakdown of members and their coronary risk factors based on their homocysteine levels.

Homocysteine levels and risk of coronary artery disease

Risk factor	Homocysteine units (micromoles per litre of blood)	Percentage of members supplementing with B vitamins in this range
Lowest risk	0–6.3	38%
Moderate risk	6.3–10	52%
Highest risk	Over 10	10%

It should be emphasised that Life Extension Foundation members are highly health-conscious individuals, for the most part eating healthy diets and taking supplements on a regular basis. This is borne out by their homocysteine levels – below 10 units for 90 per cent of them. In comparison, the average homocysteine level in both Britain and the US is approximately 10 to 11 units, according to the most recent surveys.[7]

The fact that 90 per cent of the Life Extension Foundation members had lower than average H scores indicates that eating healthily and taking B vitamin supplements will help to keep your homocysteine down. But it also shows that many are simply not taking enough of the right nutrients to get their homocysteine level into the superhealthy range, below 6.3, at which risk of disease and premature death become negligible.

■ The chances are your H score is high

Stated bluntly, we believe that the only safe homocysteine level is below 6 units. And while your score is probably higher than this right now, just a few months of the programme outlined in this book will bring it within a safe range. If your score is too high but average, meaning around 10, then you may achieve the goal – an H score of 6 or less, in just two months. It's only if your H score is extremely high, 15 or over, that it's going to take you longer.

Homocysteine is high all over the world

In Chile, at least 10 to 20 per cent of so-called healthy people tested have high homocysteine levels.[8] In Argentina, 70 per cent of the elderly have high homocysteine levels.[9] The majority of Americans have homocysteine levels above 10 units.[10] In Canada, 44 per cent of those with heart disease have high homocysteine levels,[11] compared to about 10 per cent of the general population.

If you're an Asian-American, it's even worse. One in four Asian men living in the US have high homocysteine levels.[12] In Japan almost 40 per cent of heart disease and diabetes patients have high levels.[13]

In Britain the average is in the range of 10 to 12 units. In Germany, 8 per cent of people have very high levels, above 15 units.[14] In Spain half the population have raised homocysteine, averaging 12 units.[15] Israeli men and women are worse off than Americans.[16] In Puerto Rico and the Dominican Republic, 40 per cent of the elderly have high homocysteine levels.[17] In Beijing, China, 15 per cent of people aged 35 to 64 have high homocysteine levels.[18] It's especially high in Aborigines in Australia.[19]

So high homocysteine knows no boundaries. Nor does a pattern of killer diseases, from cancer to strokes to diabetes to heart disease, which are spreading across the globe. This strongly suggests a basic underlying mechanism. We're convinced that this pattern is explained in large part by high levels of homocysteine and its concomitant methylation deficiency.

On current estimates, only one in ten people are likely to have a homocysteine score below 6.3, indicating that most people are deficient in B vitamins, SAMe and glutathione or are poor methylators and, consequently, prone to the common 21st-century pattern of diseases. If you don't like these odds and want to be in the 'superhealthy' category, read on.

5

The homocysteine scale – from sick to super-healthy

WE BELIEVE THERE EXISTS for all of us the tangible and achievable experience of a profound, lasting sense of wellbeing. That is, a consistent, clear and high level of energy, emotional balance, a sharp mind, a desire to maintain physical fitness and a direct awareness of what suits our bodies, what enhances our physical, mental and emotional health, and what our needs are at any given moment. This state of health includes resilience to infectious diseases, autoimmune diseases and allergies, and protection from the major killer diseases such as heart disease, strokes, cancer, diabetes and Alzheimer's. It consequently means a slowing down of the ageing process and living a long and healthy life with an active mind. At its most profound level, health is not merely the absence of pain, fatigue, depression or tension, but a joy in living and a real appreciation of what it is to have a healthy body and mind with which to taste the many pleasures of this world.

■ Check your health balance

Imagine that you are born with a health reserve – a certain amount of money in your health deposit account. Depending on what you eat, drink, breathe and think, money is gradually draining out of it. If you become overdrawn, your energy is low, you can't get out of bed in the morning and you'll suffer from niggling health problems, from recurring colds to persistent aches and pains. As your overdraft grows, you develop long-term diseases and when you exceed your overdraft limit, you die.

But as we'll show, you can get a regular statement on your health bank balance, and know how to increase it – by knowing your homocysteine level.

Of course, it's only at the stage of chronic symptoms and disease that conventional medicine kicks in. Once you are horizontally ill and bedbound, their job is to get you vertical again, almost always by treating the symptoms of disease, not the underlying causes. So you'll be functioning but not necessarily healthy. Most people are walking around vertically ill – standing up, but hardly bounding about full of the joys of spring.

How is your health account?

Superhealth	Vertically ill	Horizontally ill
boundless energy	constant tiredness	chronic fatigue
perspective on life	drained	exhausted
sharp mind	low concentration	constant aches and pains
positive outlook	mood swings	chronic depression
joie de vivre	exhausted by exercise	pessimism
physically fit	unfit	unable to exercise
rarely/never ill	rundown, frequently ill	incapacitated by illness
full life	easily overwhelmed	'Life is hard work'
toned body	flabby	'Life is against me'
contented	dissatisfied	despairing

Take a look at the three columns above. Where are you? A great number of people fall into the 'vertically ill' category – lacking enthusiasm and energy for life. How would it be to sit comfortably and consistently in the superhealthy category – full of energy in both mind and body?

That's the purpose of this book: to get you feeling great and living a longer and healthier life, free from preventable diseases.

■ Your H score is your secret to health

Instead of measuring whether you have a disease, the science of superhealth starts by measuring how well you are functioning – whether you are firing on all cylinders – even before the signs of disease become apparent. The objective is to improve your health status before chronic disease develops. This approach is known as 'functional medicine' and focuses on the early detection and correction of measurable, testable imbalances as a means to understand, prevent and reverse disease. The best single test is your homocysteine level, which as we've seen is a superb reflection of your B vitamin, zinc, methylation, SAMe and glutathione levels as well. From this alone you can get a clear indication of what you need to do to be superhealthy.

The homocysteine scale

Below 6   You are in the superhealthy zone, along with about 10% of the population. Well done!

6 to 8.9   You are in better than average health; however, your risk for disease is not zero. About 35% of people are in this range. Good but could be better.

9 to 12   You are 'vertically ill', experiencing average poor health with a moderate, but significant risk of premature death from preventable diseases. About 20% of all people tested are in this range.

12 to 15   You are experiencing worse than average poor health with a real risk of premature death from preventable diseases, along with 20% of the population.

15 to 20   You are in the very high risk category, with more than a 50% chance of heart attack, stroke, cancer or Alzheimer's in the next 10 to 30 years. About 10% of people are in this category.

Over 20   You have extremely significant risk, right now, of one of the five major killers or any of the 100 homo-cysteine-related diseases, and, if not already there, are heading fast for 'horizontal illness'. You are in the top 5% of 'high-risk' individuals.

Don't be upset if your H score is over 6 units. We'll tell you exactly what you need to do to get yourself into the superhealthy range. That's what Parts 4 and 5 are all about.

The next two chapters explain how to measure your homocysteine level and exactly who is at risk.

6

The 20 known risk
factors – test yourself

Our STRONG, UNEQUIVOCAL recommendation is
to have a homocysteine blood test. We'll tell you how in Chapter
7 (see page 61). It is the only sure-fire way to know where you
stand. However, below you'll see what kind of people, diets and
lifestyles generally indicate the greatest risk for having a high H
score – and there are a few surprises in store!

▪ The H signs and symptoms – check
yourself out

If you have five or more of these symptoms, it's almost a certainty
that your homocysteine is moderately to very high (9 to 15, if not
higher).

Are you tired a lot of the time?

Is your stamina, or ability to keep going, noticeably decreasing?

Are you having a hard time keeping your weight stable?

Do you often experience physical pain, be it arthritis, muscle aches or migraines?

Do you get frequent colds?

Is your eyesight deteriorating?

Is your mental clarity or concentration decreasing?

Are you experiencing more sleeping problems?

Is your memory on the decline?

Are you often depressed?

Do you average two or more alcoholic beverages daily?

Do you drink more than three cups of coffee daily?

Do you smoke cigarettes?

Are you a strict vegetarian?

Do you eat red meat at least once a day?

■ Homocysteine can be inherited

All of us inherit strengths and weaknesses. On a biochemical level, what this often means is that certain enzymes may work better than others due to our genetic inheritance. While you can't change your inherited weakness, you can maximise how well your enzymes work by consuming optimal amounts of co-factors, which are usually vitamins and/or minerals that get the most miles per gallon from your enzymes. That's why children with homocystinuria – a usually fatal genetic condition associated with very high blood homocysteine levels – do much better when given lots of vitamin B6 because this is the co-factor in their weak enzyme.

As we saw in Part 1, homocystinuria is very rare. Much more common is a genetic defect in which a particular enzyme, nicknamed MTHFR (methylenetetrahydrofolate reductase), doesn't work so well. This defect creates a logjam in the dance of methyl groups and a corresponding increase in need for B vitamins, especially folate, B12, B6 and B2, resulting in high homocysteine levels. The odds of you having this genetic twist are surprisingly high, at 10 to 15 per cent.[20] It is much more common in people of Caucasian or Japanese origin than in Africans.[21] If you have this enzymatic defect, you will need much more B12, folate, B6, B2 and perhaps other homocysteine-lowering nutrients, to keep your H score in the safe range.

How do you know if you do need more supplemental nutrients? The only foolproof way is to test your H score, and if it's too high, increase your nutrient intake accordingly (as explained in Parts 4 and 5). Then after two months of therapy you'll need to test again. However, if you have any of the homocysteine-related diseases covered in Part 3, or a family history of these diseases, your odds of having an MTHFR deficiency, and subsequently having high homocysteine, are much greater.

Check out your family history and genetics below.

Has your first-degree family (mother, father, brothers and sisters) suffered from any of the following?

- [ ] Heart disease, especially before 50 years of age
- [ ] Strokes
- [ ] Alzheimer's disease
- [ ] Abnormal blood clots

☐ Cancer

☐ Severe depression (especially in women)

☐ Elevated homocysteine levels

If the answer to any of these is 'yes', there is an increased risk that you may have inherited the same enzyme deficiencies that make you more prone to these diseases and to high homocysteine. Your need for nutrients may be massively higher than the average.

Of course, if your first-degree family doesn't have any of these but your second-degree family (grandparents, uncles, aunts, cousins) do, don't rule out the possibility that you may still have a genetic predisposition to high homocysteine.

▪ The 23 most common risk factors for high homocysteine

Apart from inheriting enzyme deficiencies, having several H symptoms or genetic predisposition to any of the medical conditions listed above, there are a number of other risk factors that increase your chances of having a high H score:

☐ Are you male?

☐ Are you aged over 40?

☐ Are you post-menopausal?

☐ Are you a regular alcohol consumer?

☐ Are you an excessive alcohol consumer?

☐ Do you smoke?

☐ Are you often aggressive or angry, or do you regularly suppress anger?

☐ Do you rarely exercise?

☐ Do you rely on stimulant drinks – tea, coffee, caffeinated drinks?

☐ Are you pregnant?

☐ Are you vegan (strict vegetarian)?

☐ Do you eat red meat or other animal protein every day?

☐ Do you think you have a relatively high-fat diet?

☐ Do you frequently put salt on your food?

☐ Do you rarely take supplements?

☐ If you do, are you just supplementing to '100% RDA' levels?

☐ Do you have an underactive thyroid (hypothyroidism)?

☐ Do you have chronic kidney problems, or a high blood creatinine level?

☐ Are you on anti-epilepsy medication?

☐ Do you supplement with more than 1g of niacin (a form of B3) daily?

☐ Do you supplement with more than 2 to 4g of vitamin C daily?

☐ Have you recently intentionally lost weight by restricting calories?

☐ Have you suffered from any of the following diseases?

- Inflammatory bowel diseases (coeliac disease, Crohn's disease, ulcerative colitis)

- Thyroid problems

- Cardiovascular disease including heart attack, angina, heart surgery, abnormal blood clotting (thrombosis)

- Strokes

- Cancer

- Diabetes

- Dementia/senility

- Osteoporosis or decreasing bone mass density

- Menopausal problems, such as excessive hot flushes

- Pregnancy problems, including repeated miscarriages, premature births, difficulty conceiving, or giving birth to an infant with spina bifida, cleft palate, club foot or Down's syndrome

- Inflammatory problems such as arthritis, asthma or eczema

- Fibromyalgia, especially with chronic fatigue

- Headaches, including migraines

- Poor kidney function or failure

- Stomach ulcers, especially if *Helicobacter pylori* infection-related

- Oral dysplasia (precancerous lesions of the mouth)

- Vaginal dysplasia (precancerous lesions of the cervix)

- Deficiency in folate, B12, B6, B2, zinc or magnesium.

The more questions you answer 'yes' to, the greater your risk of having a high homocysteine level. But the only way to find out for sure is to have a homocysteine test.

We'll be exploring the science behind the reasons why positive answers to these questions increase your risk in Parts 4 and 5.

7

How to measure your homocysteine level

UNTIL VERY RECENTLY, unless you were selected to take part in a research study, the chances of getting anyone in the medical profession to test your homocysteine level were pretty slim. Now that homocysteine is becoming more widely accepted as a marker for many diseases, this is changing – but at the moment, you're only likely to get a test if you're already sick. So if you have not been diagnosed with a disease, but want to be tested, what should you do?

You have three choices: find an informed doctor; see a nutrition therapist; or get the test done yourself. If you take the latter course, you can either visit a clinic or laboratory that offers routine homocysteine testing, or get a do-it-yourself test to do at home. (See the Resources section on page 260 for a list of the labs that do the testing, or provide and analyse DIY tests.)

When you have the test done by a doctor or a laboratory, you will need to have a blood sample taken from a vein in your arm, wrist or back of the hand. The plasma (liquid) portion of the blood is then separated out and sent to a lab for analysis. But this test can be expensive because it requires a doctor or someone trained to draw blood and process your blood sample with speed and accuracy.

The easiest – and probably cheapest – way to test your homocysteine level is with a DIY kit, which you can order yourself. Many nutrition consultants are now using this kind of testing. YorkTest, a bio-science laboratory based in York, England, has pioneered DIY home testing. At the time of publishing, this is the only laboratory we know of that offers such a test.

With the YorkTest DIY kit, you take your own blood sample for homocysteine after an overnight fast. You take a simple finger pinprick of blood, which you put on to a card that immediately separates out the plasma from your blood cells. The card with your plasma sample is then sent back to the laboratory and, within days, you will know your H score. This revolutionary approach is made possible by the unique way in which very small amounts of blood can be collected by the patient without the need for an intermediary or any fancy equipment.

The biggest heart patient group in Britain, the British Cardiac Patients Association, has unequivocally welcomed this development. An association spokesman has said, 'It is vitally important that people are informed about all of the possible risk factors of heart disease. The evidence emerging around the world about the risk of high homocysteine levels is overwhelming. However, information must go hand in hand with the availability of testing, whether at the local pharmacy or at home, and we welcome this new initiative.'

■ What happens if your level is high?

As we've said, it will generally take at least two months to bring a high homocysteine level down to a safe level, by following the programme outlined later in this book. We will tell you what to

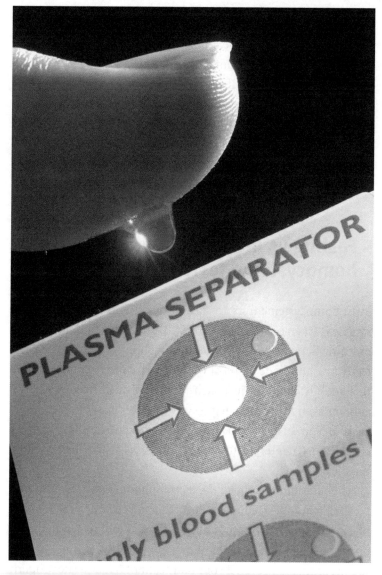

Figure 6. A do-it-yourself test for homocysteine

do whatever level your homocysteine is. The higher your level, the greater the amount of homocysteine-lowering nutrients you need.

Retesting your homocysteine

If your level is high (above 6 units) it's very important that after some time you retest to confirm that what you are doing is working. As a rule of thumb, if your level is above 9, retest after three months. If your level is less than 9 but more than 6, retest after six months. Remember, for superhealth, your goal should be to get your H score below 6, and then keep it there. Parts 4 and 5 tell you how to adjust your supplement programme as your H score moves towards the healthy range.

■ Fasting vs methionine-loading homocysteine tests

Most homocysteine tests are 'fasting' tests. This means the blood sample is taken on an empty stomach, preferably first thing in the morning before you eat. While a fasting homocysteine test is perfectly fine and, indeed, most of the research we've referred to in this book is based on just such measures, there may well be yet another even more informative way to test homocysteine – especially if you suspect you are at 'high risk' after filling out the checklist in Chapter 6, but your results come back as being within the normal range.

This alternative is called the methionine-loading test. You may recall that homocysteine is made in the body through the removal of a methyl group from the amino acid methionine. Normally the body quickly 'remethylates' homocysteine by adding back a methyl group. However, if your body cannot remethylate properly due, say, to insufficient or sub-optimal 'big four' B vitamins, you are left with dangerously high homocysteine.

You can test your ability to remethylate homocysteine by quickly supplementing a large amount of pure methionine (available from healthfood shops) four hours before taking a blood sample to measure your homocysteine levels. By doing this, you are forcing your body to prove that it can rapidly remethylate homocysteine. If it fails, your homocysteine level will remain very high (above 39 units). This indicates that you have a real problem metabolising homocysteine and is considered by many to be an indication for homocysteine-lowering nutritional therapy.

If you are already in the high-risk category, perhaps with a personal history of strokes or heart attacks, but your fasting homocysteine level came back in a relatively safe range, your doctor or nutritional therapist might recommend this methionine-loading homocysteine test just to be sure.

To do the test, you'll need to take 100mg of pure powder or encapsulated methionine per 1kg (2.2lb) of your body weight. So, if you weight 50kg (110lb, or about 7.5 stone), you need to take 100mg × 50, so that's 5,000mg of methionine, four hours before you take a homocysteine test.

Many researchers feel that this makes the homocysteine test even more sensitive and more predictive. Cardiologists from America's 1st University Clinic of Surgery tested both methods of homocysteine evaluation on a group of 85 patients with cardiovascular disease.[22] Without methionine-loading they identified 19 patients (22 per cent of those tested) with very high homocysteine scores above 15 units. By using the methionine-loading test, however, they identified a further 13 patients (another 15 per cent) with homocysteine problems. In total, they found that 37 per cent of these patients had a homocysteine problem, compared to 22 per cent with just the fasting test. This shows just how common abnormal, unhealthy homocysteine metabolism really is. Similar results have been found with stroke patients.[23] Of course, if they had used an H score of 9 units, or as we recommend, 6 units, they may have identified all 'at risk' patients without needing to do the methionine-loading test.

Within medical circles there is some debate about the value of methionine-loading. Some researchers have suggested that up to 40 per cent of those 'at risk' of unhealthy homocysteine metabolism can be missed by using the fasting test alone.[24] However, this does all depend on what level of homocysteine you consider too high. We predict that the current convention to consider a level above 12 units as high will gravitate down towards a level of 9 and then to 6.

If you do choose to have a methionine-loading homocysteine test, bear in mind that the so-called 'normal' ranges will be different. We estimate the following levels of risk, based on your methionine-loading homocysteine scores:

Risk category	Fasting H scores	Methionine-loading H scores
Superhealth	below 6	below 9
Lower risk	6 to 8.9	9 to 39
Moderate risk	9 to 12	above 39
High risk	12 to 15	60 to 100
Very high risk	15 to 20	above 100
Extremely high	above 20	above 200

If a methionine-loading test puts you into one of these 'at risk' categories, follow our advice for the category of risk given in Parts 4 and 5.

Generally speaking, if you use our recommended level of anything above 6 indicating high homocysteine, we think that the further sensitivity provided by a methione-loading homocysteine test is likely to be minimal and therefore probably not necessary, and certainly more complicated.

PART 3

50 Diseases Linked to Homocysteine

8

How to live to a healthy 100

IS A LONG AND HAPPY LIFE the ultimate pipe dream? Hardly. It's well within your grasp – and the best way to do it is not to die prematurely from chronic disease. At least three in four people die from a disease that's largely preventable. In the main, this means the big five – heart attacks, strokes, cancer, the complications of diabetes and Alzheimer's disease.

If you can prevent all of these diseases – and your chances of doing just that are excellent if you follow the H Factor programme – you are likely to add between ten and 20 years to your lifespan. If you live in the UK, the average lifespan is 73 if you're a man, and 79 if you're a woman. If you're American, the figures are 76 for a man and 83 for a woman. So an extra decade or two of healthy living would be quite a boon. And probably the best single measure that you are on the right track is, as we've now seen, your homocysteine level.

Earlier we told you about the extraordinary findings from a comprehensive research study at the University of Bergen in Norway, published in 2001.[1] They measured the homocysteine levels of 4,766 men and women, aged 65 to 67 in 1992, and then recorded any deaths over the next five years, during which 162 men and 97 women died. They then looked at the risk of death in relation to their homocysteine levels. They discovered that 'a strong relation was found between homocysteine and *all* causes of mortality.' In other words, homocysteine is an accurate predictor of death, whatever the cause.

◾ Lowering homocysteine halves your risk of dying

What they found was that the chances of a person of 65 to 67 years dying from any cause increased by almost 50 per cent for every 5-unit increase in homocysteine! This strongly reflects how central homocysteine and methylation (and, as a consequence, your B vitamin, SAMe, nitric oxide and glutathione status) are to the underlying causes of the common diseases that kill most of us prematurely in the 21st century. Turn this amazing finding the other way round and we can say that, for every 5-unit drop in your H score, you halve your risk of dying prematurely from all common causes. If, for example, your H score was 15, and you drop it to 6 units or less and maintain it there, you can probably add around ten years to your life! And it would be a lively decade because, as you will see, if you lower your homocysteine level below 6 units with the proper balance of diet and supplementation, and alleviate your methylation problems, your cells will age slower, you'll have increased vitality, and you'll feel younger than your years.

As we've noted, what we have found is that there are over 50 medical conditions associated with high homocysteine. The most common conditions linked with a high H score are shown in the chart below, with a more comprehensive list given in Appendix 2.

An A–Z of medical conditions associated with high homocysteine

Accelerated ageing

Alcoholism (and increased risk of having alcoholic withdrawal seizures)

Alzheimer's disease

Anaemias (when related to B-vitamin deficiencies)

Angina (chest pain related to coronary artery disease)

Arthritis (both osteoarthritis and rheumatoid)

Atherosclerosis

Autoimmune diseases (such as insulin-dependent diabetes, ankylosing spondylitis, rheumatoid arthritis, underactive thyroid, Hashimoto's thyroiditis)

Birth defects (such as cleft palate, club foot, neural tube defects, premature birth, urinary tract abnormalities, heart defects, pyloric stenosis)

Brain wasting or shrinking in 'normal, healthy' elderly patients

Breast cancer

Cancers generally (including cancers of the colon, thyroid, leukaemias, skin)

Cholesterol, elevated (high total cholesterol and/or LDL cholesterol)

Chronic fatigue syndrome (often associated with the pain and tenderness of fibromyalgia)

Cirrhosis or fibrosis of the liver (including alcoholic cirrhosis)

Coeliac disease

Colon cancer

Crohn's disease (with homocysteine levels often well above 40 units)

Deep vein thrombosis (abnormal blood clotting)

Dementia

Depression, severe (especially in women)

Diabetes (non-insulin dependent and insulin dependent)

Down's syndrome (high homocysteine is found in mothers of Down's syndrome children; Down's syndrome children, on the other hand, have low homocysteine!)

Epilepsy in children and adults

Fibromyalgia (especially when associated with chronic fatigue)

Folate deficiency (which can lead to anaemia, anxiety, poor memory, stomach pains and depression, cervical dysplasia, spina bifida and pregnancy problems)

Glutathione deficiency in liver, brain and elsewhere (which accelerates ageing and onset of Alzheimer's, damages liver, increases risk of stress-induced stomach ulcers, and is associated with sleep disorders, cataracts, allergies, addictions, AIDS and cancers of the lung, prostate, skin, bladder and liver)

Heart abnormalities (when present at time of birth, that is, congenital heart defects)

Heart attacks (myocardial infarction)

HIV/AIDS (the speed at which the virus develops)

Hypothyroidism (especially autoimmune thyroid disease, Hashimoto's thyroiditis)

Kidney failure, chronic/kidney dialysis patients

Leukaemias

Memory decline in older age

Mental retardation

Migraine headaches

Miscarriages

Neural tube defects in babies (such as spina bifida, anencephaly)

Oestrogen deficiency and post-menopausal symptoms

Osteoporosis

Parkinson's disease

Penile erection dysfunction

Polycystic ovary disease

Pregnancy problems (such as spontaneously recurring miscarriages, pre-eclampsia, gestosis, placental calcification, premature birth, neural tube defects)

Premature death from all common causes

Psoriasis

Pulmonary embolism (blood clots lodged in lungs)

Rheumatoid arthritis

Schizophrenia

Sleep apnoea accompanied by cardiovascular disease (sleep apnoea is also associated with snoring, temporary cessation of breathing, obesity, premature death)

Sperm motility reduction (associated with infertility)

Strokes

Thyroid cancer

Thyroid disorders (thyroiditis, hypothyroidism, Hashimoto's thyroiditis)

Ulcerative colitis (often associated with homocysteine levels above 40 units)

Vasospasm, cerebral (brain) arteries, causing strokes (often associated with magnesium deficiency)

Vasospasm, coronary (arteries of heart go into spasm, causing heart attacks, abnormal heart rhythms and premature death)Vitamin B12 (cobalamin) deficiency (which can lead to eczema or dermatitis, anxiety, lack of energy, poor hair condition, anaemia, asthma and calcium deposits, for example)

Vitamin B6 (pyridoxine) deficiency (which can lead to depression, nervousness, lack of energy, water retention, kidney stones, hyperactivity and depression, for example)

Vitamin B2 (riboflavin) deficiency (which can lead to burning or gritty eyes, sore tongue, split nails and cracked lips, for example)

Zinc deficiency (which can lead to frequent infections, slow healing, short stature, poor dark/light adaptation, lack of energy, low fertility, stretch marks, acne, loss of appetite and poor digestion, for example)

We will be discussing many of these conditions in the chapters that follow.

But why, you might wonder, is homocysteine central to so many medical conditions and diseases and such an amazing predictor of lifespan? The answer is starting to emerge from extraordinary research into how our body cells actually age.

■ Why your rate of ageing is determined by your methyl IQ

Your body is made out of well over 10 trillion cells. Some cells live only a few days, others for several months and a few live for years, then they die and are replaced. While reading this book today, for example, millions of your immune cells and cells lining your small intestine will die and be replaced with new cells.

These cells are built using a blueprint, which is contained in each cell's DNA. So, the ability to build a healthy cell depends on the state of that DNA map. If the map has been photocopied and reproduced dozens of times, the way your cells are reproduced through your lifetime, it gets a bit blurry in places and hard to read – and signs of ageing and disease start to appear.

Or you can think of your DNA as a huge castle. Each room represents a different set of instructions that, taken all together, create the trillions of cells that make you complete. One wing might be the instructions for building a liver cell, and one room within that wing might be the instruction for constructing an enzyme that detoxifies alcohol or homocysteine. Each individual cell carries out a particular function – so the instructions are tailored to that function only. This means that in any given cell, most of the DNA is inaccessible – the doors to these rooms are shut.

But it is possible to open these doors: and the secret to this selective access is a substance called chromatin, a protein/DNA complex that makes up about half the space around DNA. We used to think that chromatin was like packaging, but more recent research points suggests that chromatin controls access to information contained in the DNA. It holds the keys that lock or unlock the rooms.

These keys are called histones. They are like arms that reach into the DNA. It is here that homocysteine and the methylation process enter the picture. Whether the key turns or not depends on methylation. If there's a methyl group on the end of a histone, the room stays locked, and if the methyl group is removed, the room opens. This is a major discovery because what it shows is that it isn't just the DNA that's important, but the ability to access the right information at the right time – and this depends in large part on how well you can move your methyl groups around. And remember: that's exactly what homocysteine measures – your methyl IQ. The lower your homocysteine, the higher your methyl IQ, and the younger and healthier your cells remain.

■ High homocysteine makes your cells age quicker

Of course, your DNA doesn't really resemble a castle at all. It's packaged in chromosomes within the nucleus of your cells. And there's a way of measuring how battered your DNA has become over the years. At the end of each chromosome is something called a telomere that protects it from damage, in much the same way that the hard plastic bit at the end of a shoelace protects it from fraying.

The more often we replace cells, the more damaged and the shorter the telomere becomes. If the telomere becomes quite worn down, the DNA in the chromosome is more exposed, not only to damage, but to being read wrongly. So, the search for the culprits of ageing and disease is a search for what shortens telomeres. And the answer is in large part homocysteine.

Let's make this real. If you expose the cells which line arteries throughout your body to homocysteine, the cells get older much quicker.[2] If you look close up, what you would find is that as these cells die and are replaced, the telomere on the end of each chromosome gets shorter and shorter. This means that the cells are ageing rapidly and are more prone to damage and premature death. And the main cause of this damage is oxidation (see page 30).

A lack of methyl donors and movers, which is precisely what a high homocysteine level indicates, also means less accurate access to the right pieces of building instruction. Since high homocysteine also means you have less glutathione, this also means more bombardment by oxidants. In a nutshell, high homocysteine adds years to your cells, while low homocysteine takes years away.

■ Why homocysteine ages your body

It's well known that the cells that line the arteries of people with heart disease look like much older cells. Professor Nilesh Samani and his team from Leicester University in the UK found that the average cells of a person with heart disease are nine years older than those of healthy people. In essence, these people were nine years older than their biological age.[3]

Looking closely at these artery cells, the team found that the telomeres – perhaps the best measure of the cell's biological age – were much shorter in those with heart disease, the most likely cause being high homocysteine.

Other researchers have gone on to investigate what causes the damage to our chromosomes and DNA and leads to accelerated ageing and disease. Research by Dr Ann Fenech, published by the New York Academy of Sciences, found that a high level of chromosomal damage was more prevalent in those with low folate or B12 status, and therefore high homocysteine. From this they were able to work out the ideal blood homocysteine level for minimising chromosomal damage. Keeping in mind that the average H score in the general British and American population approaches 11 units, in line with our thinking Fenech's research sets the optimal homocysteine level at less than 7.5 units.[4]

Finally, we are discovering that the reason we age is twofold. Cells, and their replacement instructions, get increasingly damaged by poor methylation and excessive oxidation, both of which are reflected in your H score. By following the H Factor diet and supplement programme, you can slow down the very process of ageing and achieve your full biological potential, which, for most of us, should mean living free of disease beyond 100 years.

9

Halve your risk of a heart attack

HERE'S ONE OF THOSE good news/bad news stories. First the bad news. Almost one in two men *and* women die from heart attacks or strokes. One man in every four will have a heart attack before retirement age and a quarter of all deaths from heart attack occur in people under the age of 65. For women, heart disease and strokes are second only to cancer as the leading cause of death between the ages of 35 and 54.

While heart disease usually strikes after the age of 45, even by the age of ten fatty deposits, which mark the beginning of arterial disease, are already present in most Western children's arteries. So widespread is this modern epidemic that we almost take heart disease for granted. We fail to protect ourselves from a condition that is not only thoroughly studied, but eminently preventable.

There is nothing natural about dying from heart disease. Many cultures in the world do not experience a particularly high incidence of strokes or heart attacks. Americans and Britons, for example, have approximately nine times as much heart disease by middle age as the Japanese, although this difference is starting to narrow as the Japanese adopt more Western diets and lifestyles. But it wasn't always this way.

Autopsies performed on mummified Egyptians who died in 3000 BC show signs of deposits in the arteries but no actual blockages that would result in a stroke or heart attack. Despite the highly distinctive signs and symptoms of hearts attacks (severe chest pressure and pain, cold sweats, nausea, fall in blood pressure, weak pulse or sudden death), in the 1920s they were so rare that it took a specialist to make the diagnosis. According to American health records, the incidence per 100,000 people of heart attacks was near zero in 1890. Although deaths did occur from other forms of heart disease, including calcified valves, rheumatic heart and other congenital defects, the incidence of actual blockages in the arteries which cause strokes or heart attacks was minimal.

By 1970, however, that incidence had risen to 340 deaths per 100,000. Today, 44 per cent of all deaths in the US are caused by heart attacks. In Britain, a quarter of a million people die prematurely every year from heart attack or stroke.

Even more disconcerting is the fact that heart disease is occurring in younger and younger people. Autopsies performed in Vietnam showed that one in two American soldiers killed in action, whose average age at death was 22, already had significant atherosclerosis (a build-up of plaques or tumour-like deposits in the walls of arteries that interfere with blood flow). Nowadays most teenagers can be expected to show signs of atherosclerosis, which heralds the beginning of heart disease. Obviously, something about our lifestyle, diet and/or environment has changed radically in the last 70 years to bring on this modern epidemic.

The cost of heart disease is, on average, a life shortened by 20 years. While the healthy human lifespan is a maximum of at least 100 years, the vast majority of people are dying prematurely from heart disease, strokes or cancer. Even those who do reach their nineties are often not in good health, suffering from Alzheimer's, complications of diabetes, extreme fatigue, chronic depression and/or crippling arthritis. No wonder we sometimes ask, 'Who wants to live that long anyway?'

▪ Homocysteine and cardiovascular disease

Cardiovascular disease, which is popularly called heart disease, is a blanket term for a number of diseases of the blood vessels. These include heart attack, heart failure, angina pectoris (chest pain or heavy pressure from poor blood circulation to the heart), stroke and peripheral vascular disease (blood supply is restricted, for example to the limbs, as a result of clogged-up arteries).

If you have cardiovascular disease, or have a history of strokes or heart attacks, the chances that you have an unacceptably high H score (over 9 units) are well above 50 per cent. About 30 per cent of you will have a level above 15 units, which we classify as very high. Very conservatively, we estimate that at least 8 million people in Britain and 40 million Americans have dangerously high homocysteine, increasing their risk of a heart attack by at least 50 per cent.

The higher your H score, the higher your risk. To put this in perspective, every 12 per cent increase in your H score triples your risk of a heart attack, if you are a man. If you have both a high H score and a family history of heart disease, this increases your personal risk of heart attack, regardless of gender, by a whopping 13 to 14-fold!

The largest review of all homocysteine research to date, published in the *British Medical Journal*, confirms this strong association. David Wald and colleagues from the Department of Cardiology at Southampton General Hospital reviewed 92 studies that measured homocysteine in more than 20,000 people. Since over 10 per cent of the population have a genetic mutation that increases homocysteine levels (see page 57 for more on this), the report split the groups into those with or without this gene mutation.

They found that with every 5-unit increase in homocysteine measured in the blood, the risk for heart disease went up 42 per cent in those with the gene mutation and 32 per cent in those without. The risk for stroke went up a massive 65 per cent in those with the genetic mutation and 59 per cent for those without. The researchers concluded that these 'highly significant results indicate strong evidence that the association between homocysteine and cardiovascular disease is causal'.[5] This means that having a high homocysteine level isn't just associated with higher risk, it actually causes heart disease – a conclusion that is also being reached by other research groups.[6] This means that if you can lower your H score you remove the cause, and hence a major risk.

According to this study, this means that lowering a high homocysteine level from 15 to 6 units and maintaining it there might cut risk by a breathtaking 75 per cent! This is not only much more substantial a reduction than from lowering cholesterol levels, it's also more easily achievable. How? With nutritional supplements and dietary change, and certainly not drugs.

Move over cholesterol

Cholesterol may have been hogging the spotlight in the diagnosis of heart disease risk, but the fact is that your H score is roughly 40 times more predictive of a heart attack! Time and time again we encounter heart attack patients who don't have high cholesterol levels, haven't been checked for homocysteine and are still put on cholesterol-lowering drugs such as statins. While these drugs do lower deaths from heart attacks, they don't have as great an effect on overall mortality, adding on average only 18 months to life. Lowering your H score, on the other hand, dramatically reduces the risk of death from all causes, not just heart attacks.

If you are at risk of heart disease, measuring your cholesterol but not your H score is a potentially fatal mistake. If your H score is high, taking steps to lower it will most likely improve your cholesterol status as well. This is because too much homocysteine damages cholesterol through oxidation, making it much more dangerous.

Heart surgery patients with high H levels have many more problems

Every year in Britain, 28,000 people have coronary bypass surgery (300,000 in the US), grafting on blood vessels taken from elsewhere in the body to bypass the blocked coronary arteries. During the first year after bypass surgery, up to 15 per cent of these blood vessel grafts begin to fail, becoming blocked all over again. After 10 years, atherosclerotic clogging in the grafted blood vessel develops in up to 30 per cent of all cases.

Comparing homocysteine levels of patients who have had heart transplants with healthy controls has demonstrated a much higher average homocysteine level in patients (19.1 units), with two-thirds of heart transplant patients having very high H scores above 15 units. Those with high H scores were also more likely to have complications associated with the surgery.[7]

One of the most common surgical procedures for those with coronary artery disease is angioplasty. It involves inserting catheter tubing containing a small balloon into arteries around the heart. The balloon is inflated to flatten deposits of atherosclerotic plaques blocking the artery, so blood can once again flow to the heart.

Like bypass surgery, angioplasty is often not a permanent cure, and after surgery the arteries may reclog in the treated area – a very undesirable condition called restenosis. This has led to the use of alternative techniques such as laser angioplasty, in which lasers are used to burn away or vaporise the plaques.

Restenosis of the coronaries is much more likely if your H score is high, according to research at the Swiss Cardiovascular Centre in Bern.[8] In a nutshell, this means the higher your H score, the faster your coronary arteries will narrow again after surgery, so there's little point having an angioplasty without testing for and treating high homocysteine.

If heart surgery patients with high homocysteine supplement with therapeutic amounts of folate, B12, B6, B2, zinc and TMG (and magnesium – see below), they are much less likely to have another heart attack, die from heart disease, have complications from surgery and need further surgical intervention.[9]

Consider the case of Garry.

case study

Having suffered from angina for years, Garry had a bypass in April 1998. It wasn't as successful as he'd hoped as, although it had improved, he still suffered from some angina. In addition, he developed a condition called Dressler's syndrome – not uncommon after a bypass – in which the heart tissue becomes inflamed. By 2001 one of his coronary arteries was almost completely blocked, so he had an angioplasty, and a stent put in.

Garry then consulted Dr Eric Asher, who suggested that high homocysteine may be the cause of his problems. On testing in July 2002, his H score was 18, putting him in the very high risk category. Garry was given folic acid, B6 and B12 and some concentrated fruit and vegetable supplements, also high in these B vitamins. On re-testing five months later, his H score had halved to 9, significantly decreasing his risk. Garry no longer needs to take steroid drugs for the Dressler's syndrome and his angina is much better. By lowering his H score he has also dramatically reduced his risk of further atherosclerosis.

Homocysteine linked to high blood pressure

Your blood pressure is controlled by the changing diameter of your arteries – and this can be made smaller by blockages, artery muscle spasms, stiffening of arterial walls and your mineral balance. Magnesium relaxes the arterial muscles, lowering your blood pressure, while sodium (salt) raises your blood pressure. Given that arteries contain a layer of muscle, Professors Burton and Bella Altura, a husband-and-wife team from the State University of New York's Health Science Center, wondered whether a deficiency in magnesium could cause an artery to spasm, reducing or cutting off blood supply. Their research,[10] which spanned three decades, proved conclusively that removing magnesium from the environment of blood vessels made them go into spasm, potentially reducing the diameter of an artery by two-thirds. As long ago as 1977, researchers from Georgetown University in Washington, DC, demonstrated an 11 per cent decrease in blood pressure by giving magnesium to those with hypertension.[11]

While much of the research into magnesium's protective effects has focused on its relaxing role in arteries, there is growing evidence that magnesium can also reduce your levels of cholesterol and triglyceride (fats transported in the blood).[12] Indeed, having a low level of magnesium is as great a risk for cardiovascular disease as having a high cholesterol level.

The Alturas recently investigated whether having a high H score had any effect on magnesium. They found that high homocysteine caused rapid depletion of magnesium from the smooth muscles that line the arteries, making them more likely to go into spasm, which can precipitate a heart attack or stroke.[13] Interestingly, they also found that if magnesium levels were low and homocysteine high, the combination of vitamins B6, folate and B12 – the usual homocysteine-lowering cocktail – was ineffective.

If magnesium is the good mineral, sodium is the bad one. Once again, having a high H score makes matters worse. According to research in Japan, the combination of a high salt intake and high homocysteine further constricts blood vessels.[14]

So, up your daily magnesium intake by eating lots of fruit, vegetables, nuts and seeds, supplement with chelated magnesium (magnesium citrate or glycinate is a good one), and lower your salt intake. These measures, plus other supplement recommendations and dietary changes explained in Part 4, are all effective ways of lowering your H score.

If you do have high blood pressure we strongly recommend you have your H score checked because it is often high in such cases and, if so, gives you a direct means to prevent a very important underlying cause of restricted arteries.

Valda and Ian are cases in point:

case studies

Valda, aged 73, had suffered from high blood pressure for over 30 years, as well as a touch of arthritis. Her doctor had given her two drugs, Captopril and a junior aspirin, to take every day. They had helped a bit, but her blood pressure was still high – somewhere between 140/70 and 160/80. She decided to have a homocysteine test. Her H score was 42.9, putting her in the very high risk category. She is now embarking on the H Factor diet and supplement programme and will retest in a few months.

Ian, aged 56, started getting excessively tired and breathless, with frequent pounding headaches and eye pressure. He was diagnosed with high blood pressure, of around 180/120. He was put on two blood pressure lowering drugs, bendrofurozide and irbecartan, which have helped, together with

▶

some diet and lifestyle changes, to lower his blood pressure to 140/75. He drinks less coffee and has reduced the stress in his life, which was considerable as he is a journalist. When he had his homocysteine level checked he scored a massive 42.8, putting him in the very high risk category. He is now embarking on the H Factor diet and supplement programme and will retest in four months.

Say NO to heart disease

Another big factor in high blood pressure and heart attack risk is a deficiency of nitric oxide (NO) in your artery walls. As we saw on page 32, this highly versatile and important biochemical regulates the muscle tone of arteries. It also prevents atherosclerosis and injury to the vessel wall. At the same time it functions elsewhere in our bodies as a powerful antioxidant, anti-inflammatory, brain cell protector, neurotransmitter (chemical messenger in the brain) and memory and learning enhancer. Not only this: it also augments erections in men, relaxes the bladder wall and dilates the airways.

The more NO is produced in the arteries, the more the arterial muscles relax and increase in diameter. Blood pressure also falls in the process. Nowadays, many hypertensive drugs work by increasing NO activity. But guess what lowers it to undesirable levels? High levels of homocysteine in your blood.[15, 16]

Homocysteine not only lowers NO and interferes with magnesium (both of which raise blood pressure), it damages the walls of the arteries, oxidises cholesterol, and makes the blood stickier and more likely to clot abnormally. In animal studies at the University of Utah, such negative effects were induced by lowering intake of folate (a B vitamin), which then raised homocysteine levels.[17]

The four independent mechanisms for promoting heart attacks and strokes detailed above explain why having a high H score increases your risk of these potentially fatal events by almost 80 per cent. One, but by no means the only, preventive step you can take is to increase your daily intake of folate. The other steps are discussed in Parts 4 and 5.

10

Preventing strokes – the silent killer

WHILE HEART ATTACKS claim more lives, strokes strike over 750,000 Americans each year, of those killing 160,000. It is also a common killer in Britain, claiming 60,000 lives each year, which makes it the third leading cause of death behind heart disease and cancer.

One in four men and one in five women can expect to have a stroke if they live to the age of 85. Strokes often strike out of the blue and, when not fatal, leave many – both young and old people alike – with varying degrees of debilitation. There are no miracle drugs or surgical procedures that can undo the brain damage of a stroke. Therefore, any way of predicting who is heading for a stroke and taking effective steps to head it off is of paramount importance.

Interestingly, although high blood pressure has received much of the attention in stroke research, two-thirds of stroke victims do not suffer from it. Researchers are looking elsewhere for the answer. One again, the answer is homocysteine. While up to 30 per cent of heart attack victims have high homocysteine levels, at least 40 per cent of stroke victims do. In fact, your H score is just about the best predictor around. And if you can then reduce high levels, you can make a very positive step towards protecting yourself from ever suffering a stroke.

■ High homocysteine increases stroke risk by 82 per cent!

Consider a recent study involving 1,158 women and 789 men aged 60 years or older. They had taken part in earlier studies, one between 1979 and 1982 and the next in 1992, looking at homocysteine levels as a predictor for having a stroke.[18, 19] Seven years later, doctors found that those who had an H score above 14 units had an 82 per cent increased risk of total stroke, compared to those with less than 9.2 units!

This dramatic increase in risk was more significant than a stroke victim's age (each additional year adds only a 6 per cent risk), blood pressure or whether or not they smoked. All of which makes homocysteine the single most important risk factor to check for likelihood of strokes or heart attacks!

A recent review of most studies to date, published in the *British Medical Journal*, shows that risk of a stroke goes up about 60 per cent with every 5-unit increase in homocysteine.

Consider the case of John and Janet.

case study

Janet's husband John had a stroke in 1999 – and then another the following year. Among other alternative therapies he had a homocysteine test. His level was 17.9. He changed his diet and started taking the recommended vitamins, and four

▶

months later his homocysteine level had dropped to 9. Reading about the importance of homocysteine, Janet, aged 61, decided to have hers measured. It was 25.6, putting her in the very high risk category. John and Janet are now supplementing large amounts of B2, B6, B12, folic acid and TMG (betaine). They will retest again in a couple of months.

As every month goes by, more and more medical research groups are reporting similar findings. In Japan, at the Department of Medicine and Clinical Science at Kyushu University, Dr Kensuke Shimuzu and colleagues have recently reported a strong association between a person's H score and subsequent risk of a stroke.[20] Another research group in Japan, this time at the Department of Geriatric Medicine at Tohoku University School of Medicine, wondered whether homocysteine levels might be linked to 'silent strokes', commonly found in the elderly. These often go unreported but lead to a decline in mental functioning and an increased risk of death from subsequent strokes.[21] So they selected 153 elderly people over the age of 66 and gave them an MRI brain scan, from which you can diagnose a silent stroke. They found that a quarter of the participants had had a silent stroke and compared their H scores to those who hadn't. In those with H scores above 15, the risk of having a silent stroke was substantially higher than for those with lower levels. The average H scores of those who had had strokes was 13.6 units, compared to 11 units for those who hadn't.

Cardiologists the world over are gradually waking up to the fact that homocysteine is at least as important, if not more so, as a person's cholesterol level or blood pressure for predicting heart attacks or strokes. In a recent statement from the Department of Medicine at the Good Hope Hospital in Sutton Coldfield, in the Midlands, Doctors Sarkar and Lambert said, 'It [homocysteine] is at least as important as cholesterol, lipoprotein abnormalities and hypertension and should be part of risk assessment, especially those at high risk. Moderately elevated plasma homocysteine concentration is readily correctable by folate, betaine (TMG, an excellent methyl donor), or vitamin B12 supplementation.' (We will forgive them for not mentioning B6, B2, zinc and magnesium.)

■ Thrombosis and other arterial diseases

Strokes and heart attacks aren't the only diseases to strike the arteries. Less fatal but very common is peripheral artery disease (PAD), in which NO deficiency and plaque cause the arteries to become stiff and constricted. People with PAD are six times more likely to die within ten years from cardiovascular disease than those who do not have PAD. It is believed that PAD affects between 8 and 12 million people in the US and an equal number in Europe.[22]

One of the dangers associated with PAD is thrombosis or abnormal blood clots, usually found in the legs, lower abdomen and arteries to the brain. Often these blood clots break loose and are carried to the lungs or brain, where they can be deadly. Both PAD and abnormal blood clots are strongly associated with high H scores. About 30 per cent of people with PAD have high levels of homocysteine,[23] as do between 64 and 75 per cent of those with thrombosis.[24, 25] The main mechanism for the formation of these blood clots, which can occur in different parts of the body – for example, behind the knees, calves or groin – is a sticky, fibrous substance in the blood called fibrinogen. When fibrinogen levels go up, the blood starts producing blood clots. German researchers at the Institute of Epidemiology and Social Medicine at the University of Munster have also found there is a direct association between high H scores and high levels of fibrinogen, one of the best predictors of thrombosis.[26]

In summary, anyone who has any history, including a close family history, of heart disease, strokes, high blood pressure, abnormal blood clots, or other circulation problems should be routinely checked for homocysteine. If your H score is high, follow the guidelines in Parts 4 and 5 to bring your homocysteine level rapidly below 9, and then below 6, into the superhealthy range.

11

How to cut your cancer risk by a third

THE CELLS IN YOUR BODY go through three phases. First, they grow, then they do their work, then they die and are replaced by new cells. They start off 'undifferentiated', meaning that they haven't yet developed into specialised breast or liver or brain cells. But in this state they're vulnerable to cancer. Put simply, cancer is a condition where there are too many abnormal growth messages, which results in these undifferentiated cells multiplying uncontrollably and becoming cancerous. In different parts of the body these cancerous cells can start to over-multiply and take over certain organs or tissues.

While a great deal is known about the factors that contribute to cancer, from a prevention point of view one of the most critical questions is, 'What goes wrong at a cellular level? How does a "normal" cell become a cancer cell?'

One of the most promising avenues has been looking at how the cell's genetic code, the DNA, gets damaged. Often, this is triggered by oxidation, which you may remember is damage by an excess of deviant, highly reactive oxygen molecules. And whether that's brought about by an unhealthy diet, cigarette or exhaust fumes, all of these are strongly linked to lung cancer.

To put this all into perspective, up to 85 per cent of the causes of cancer are environmental – meaning that cancer is closely related to what you and I eat, drink, breathe and how we choose to live. The remaining 15 per cent is considered to be inherited, some part of which is our genetic inheritance. This is not to say, however, that 15 per cent of cancer cases are genetic. Even people with 'cancer genes', such as the BCRA1 gene for breast cancer, won't develop cancer without the environmental insults that tip the scales from healthy into abnormal cell growth.

▪ How high homocysteine and faulty methylation cause abnormal cell growth

If oxidation is one of the prime ways DNA becomes damaged, high homocysteine and associated abnormal methylation is another. DNA is always being damaged, often by oxidants, and therefore needs to be constantly repaired. It also needs to be copied, encoding new cells that we make at an extraordinary rate of tens of millions per minute. Methylation controls both the synthesis and the repair of DNA, putting homocysteine, and the key homocysteine-lowering nutrients such as B12, folate, vitamin B6, B2 and TMG, smack in the middle of the whole cancer process.

Any failure to do the right thing with methyl groups could conceivably increase the risk of abnormal cell growth. As we saw in Part 1, there are a number of nutrients involved, both the methyl donors such as TMG, and the nutrients needed for the methylation to proceed, mainly the B vitamins. Any lack of these B vitamins is already well established to increase the risk of certain cancers.

■ High homocysteine predicts cancer

Does a high H score increase your risk of cancer? This is a key question and one that is only starting to be explored. As with heart disease, having accurate markers for cancer helps in the diagnosis, prevention and treatment. Such markers can not only identify someone at risk, but can also encourage immediate preventive steps and even measure the success of a cancer treatment.

Dr L. L. Wu and colleagues at the University of Utah's Health Science Center wondered whether homocysteine might act as a tumour marker, so they decided to measure homocysteine along with other known tumour markers in cancer patients undergoing treatment.[27]

They found that when the other tumour markers went up, the homocysteine went up, and when the tumour markers went down, the homocysteine went down. They also observed that homocysteine proved to be a better marker than the other more conventional indicators. Remarkably, homocysteine also predicted much more accurately whether cancer therapy was going successfully or not. If the cancer was growing larger and therefore not responding to therapy, homocysteine increased at the same time; if the cancer was growing smaller with therapy, homocysteine levels decreased. Among the tumour markers, only homocysteine revealed success of cancer therapy in this way. Although it's in the early days of research, this study certainly indicates that homocysteine levels may prove to be a very useful indicator of the existence of cancer as well as the success or failure of cancer therapies.

■ Homocysteine, leukaemia and dysplasia

Most people have heard of leukaemia, a cancer of the bone marrow in which the number of white blood cells in the blood greatly increases. There are also a number of similar but lesser-known (and often less severe) conditions that originate in the bone marrow where the body's cells are made – and all are called 'bone marrow myelodysplasias'. Dysplasia means that the cells are malformed as if overgrowing – and this can occur not just in the bone marrow, but in any area of the body.

Leukaemia and dysplasia affect about 10,000 people, mainly children, in the US and about 2,000 in the UK. These conditions appear to be more genetic than environmental.

Recent research at the Department of Medical Science at the University of Milan in Italy has found that 42 per cent of those with myelodysplasias have high H scores.[28]

Another form of dysplasia is cervical dysplasia, a pre-cancerous lesion of the vaginal cervix. Once again, research has found that the more severe the dysplasia, the higher the homocysteine levels.[29] Conversely, the higher the folate and B12 levels, the less severe the dysplasia. These, of course, are the vitamins that lower high homocysteine levels.

So if you have a family history of cancer or dysplasia, have developed one of these conditions or are fearful you may, getting your homocysteine tested and following the H Factor diet and supplement programme to reduce it to safe levels is your first priority. This not only helps prevent cervical dysplasia, it also helps to reverse the condition if you have it.

■ Homocysteine and colon cancer

It is well known that colon cancer risk is strongly linked to a poor diet – diets high in cooked, especially burnt, meat and low in fibre, fruit and vegetables – but that taking in large amounts of folate, a key nutrient in vegetables, is highly protective.[30] This is also true for breast cancer. Could this be because the absence of folate means higher homocysteine and more methylation, oxidation and glutathione problems, leading to DNA damage? The cancer–homocysteine link gets even more interesting due to the discovery of two different defective genes that cause methylation problems. These genes code for enzymes that convert potentially toxic homocysteine into useful substances. One is the cystathionine beta-synthase enzyme (or CBS for short) that turns homocysteine into cystathionine and then on to glutathione. The other is the MTHFR enzyme that converts homocysteine into SAMe, your body's number one methyl donor. (See Figure 3 on page 25.)

Researchers in the Department of Surgery at the University of Western Australia wondered whether there might be a connection between inheriting the damaged MTHFR gene, homocysteine and colon cancer. So they checked for this genetic mutation in over 500 patients with colon cancer and found that, especially in older patients, the abnormal gene and elevated homocysteine were often present.[31]

Both genetic mutations, interestingly enough, also seem to need more vitamin B2 (riboflavin) in order to work optimally in keeping homocysteine low. This means that the combination of a damaged gene, plus low folate, which is very common in older people, and possibly low vitamin B2, leads to a high risk of disease and ill health. How do you know if you have this genetic defect? It's likely that in the not too distant future a genetic test may become commercially available, but for now, having a high H score that doesn't drop with B12, B6 and folate supplementation alone is a possible indicator. But even if this is the case, the right combination of nutrients, which should also include B2, zinc and TMG, at the right levels, can lower your H score and therefore reduce risk and reverse ill health. This is explained fully in Parts 4 and 5.

■ Homocysteine and breast cancer

The same risk combination (damaged gene, plus low folate and vitamin B2) holds for breast cancer, too. And a high intake of folate is also associated with a reduction in risk. Just as for colon cancer, a particular gene mutation, this time called COMT, leads to high homocysteine and methylation problems. So, once again, researchers wondered whether women with breast cancer might have methylation problems as reflected by high homocysteine levels, in part due to this genetic risk factor. To date research has shown that people with this gene mutation do indeed have raised homocysteine levels, but as yet, no strong association between homocysteine and breast cancer has been found.[32] However, once you know your H score, no matter what your genetic make-up, you can lower it and reduce your risk of breast cancer, or any disease associated with high homocysteine, with the right diet and supplements (again, see Parts 4 and 5).

▪ The cancer marker of tomorrow?

It is highly likely that, as the spotlight focuses on homocysteine as the marker for methylation problems, and methylation problems being seen as part of the root cause of many cancers, that we'll start to see an association between homocysteine and many different types of cancer.

For now, we can only say there is reasonable evidence that having a high homocysteine level increases cancer risk, especially colon cancer, skin cancer of the head and neck, leukaemia, and the dysplasias, including cervical dysplasia. And there is strong evidence that the H Factor diet and supplement programme will do much to cut cancer risk, probably by at least a third.

12

Diabetes and homocysteine

DIABETES IS ON THE RISE. In fact, it's one of the fastest growing diseases of the 20th and 21st centuries – and as such, a very serious problem.

The condition involves a number of serious and potentially life-threatening side-effects of having too much sugar (glucose) in the blood. Glucose is derived from the carbohydrates we eat, such as bread and fruit, and in excess it is toxic stuff. It's the body's high octane fuel and is meant to be carefully passed from the digestive tract into the blood, and into body cells where it is combusted to release energy. But in diabetes the process fails. The hormone that helps carry glucose from blood to body cells, insulin, either doesn't do its job properly (in the case of adult-onset or Type II diabetes) or just isn't there to do the job (as in juvenile-onset or Type I diabetes).

Both of these conditions are strongly linked to homocysteine.

▪ Insulin-dependent diabetes and homocysteine

In Type I diabetes, also known as insulin-dependent diabetes, mellitus or IDDM, the deficiency in insulin is usually caused by damage to insulin-producing cells in the pancreas. This damage results because the body's immune system mistakenly attacks these cells. A growing number of researchers are concluding that IDDM is sometimes caused by giving infants highly allergenic foods – particularly wheat, soya and dairy products – in the first four months of life, before their gut and immune systems are mature. There is, however, also a strong genetic component to IDDM.

In a Belgian study of 71 IDDM diabetics 23 to 63 years old, 17 per cent were found to have very high homocysteine levels, above 15 units. The average was 9.2 units, which is already a bit high. Conventional blood tests found no apparent lack of B vitamins despite the high homocysteine levels. (It is now becoming widely recognised that having a high homocysteine level is the best possible functional test for determining your folate and B12 vitamin status.) Not surprisingly, older diabetics or those who had had the disease longest tended to have the highest H scores.[33]

▪ Diabetes, sugar, bad fats, stress and stimulants

The most common form of diabetes, Type II, is the direct consequence of having high blood sugar levels too often. Usually this happens when a person eats too much sugar, refined carbohydrates and the wrong kind of fats on a regular basis, has multiple nutrient deficiencies and fails to exercise. Type II diabetes is also linked to too many stimulants, such as caffeinated drinks or smoking, and stress. Each of these – sugar, bad fats, stress and stimulants – destabilises blood sugar control, resulting in abnormal highs and lows. This, in turn, means more and more insulin is produced and, eventually, the insulin receptors on the body's cells start to become immune to it. This is called 'insulin resistance'.

At least one in four people with no diagnosed history of blood sugar problems have this condition and, if left unchecked, it can slide into diabetes. Almost all obese people, especially those with 'pot' or 'beer' bellies, have insulin resistance, which is why obesity increases a person's risk of diabetes 77 times.

Researchers from the Department of Endocrinology and Diabetes at Australia's Royal Perth Hospital investigated the link between homocysteine and both types of diabetes in 700 people. Homocysteine levels were highest in men, older diabetics and those with poor intakes of folate and B12. Not surprisingly, the strongest link was between homocysteine and those with a history of coronary artery disease and stroke.[34] These are a common complication of diabetes, partly because too much glucose in the blood damages arteries, but also quite possibly because diabetics have high homocysteine, which ups the damage considerably.

▪ Why overweight children are high homocysteine risks

Overweight children with high levels of insulin in their blood are also likely to have high levels of homocysteine, and lower levels of folate which reduces homocysteine levels, according to research at the University of Graz in Austria.[35]

The combination of elevated homocysteine and reduced folate puts these children at increased risk of developing heart disease later on. Dr Siegfried Gallistl, who led the research, said their study implied that 'the reduction of cardiovascular risk factors (such as) body fat and insulin – by dieting and/or physical activity – might improve homocysteine metabolism'. He also noted that insulin appears to inhibit enzymes that play a role in keeping homocysteine at bay.

This finding – that insulin stops the body from lowering homocysteine – is very important because we may yet discover that the high risk of heart disease and stroke associated with diabetes is also strongly linked to homocysteine. Other common complications of diabetes, such as declining mental alertness and memory, have also been linked to high H scores in diabetics.[36] High blood pressure is often present in diabetics, and we've seen that homocysteine has been linked to lowering nitric oxide in the walls of arteries (see page 32), causing the walls to become stiff and blood pressure to rise.[37]

Although no studies have yet been published, it is highly likely that a homocysteine-lowering diet and supplement programme as explained in Part 4 and 5 would work wonders for anyone with diabetes.

13

Halve your risk of Alzheimer's

WE'VE ALL HEARD about how our population is getting older. Now it's predicted that by 2030, some 20 per cent of people over 65 will have Alzheimer's. The stress and unhappiness it brings, both to the sufferer and to their family, is immense.

The first link between Alzheimer's and homocysteine was reported by British and Norwegian scientists at the Universities of Oxford and Bergen, where they discovered exceptionally high levels of homocysteine in the brain tissue of those who had died with Alzheimer's.[38] Since then, many different research groups have found high levels in those suffering with both age-related mental decline and Alzheimer's.

While these discoveries may provide an important clue to the prevention of Alzheimer's, it doesn't prove that high homocysteine actually causes Alzheimer's. But both cardiovascular disease and the poor blood circulation to the brain and silent strokes that high levels of homocysteine cause, plus low levels of B vitamins, methylation and glutathione, are also associated with increased risk of Alzheimer's. However, some researchers have argued that the high H score found in Alzheimer's sufferers is simply showing the person has these conditions, rather than proving a link.

A recent study from the School of Medicine at the University of California, Davis, is a case in point. Researchers compared 43 patients with Alzheimer's to 37 without. They found a stronger association between having a high H score (above 12 units) in those with cardiovascular disease, in both groups, than between a high H score and having Alzheimer's. They also found that those with low levels of B6 in the blood were more likely to have high homocysteine levels.[39] The researchers concluded that 'elevated plasma homocysteine in patients with Alzheimer's appears related to vascular disease and not Alzheimer's'. In addition, 'low vitamin B6 status is prevalent in patients with Alzheimer's. It remains to be determined if elevated homocysteine and/or low vitamin B6 status directly influences Alzheimer's pathogenesis or progression.'

This is one of those 'chicken or egg' situations that will take some unravelling. If homocysteine causes arterial disease, and arterial disease causes Alzheimer's, then homocysteine contributes to Alzheimer's – and most importantly, taking the steps to lower your H score reduces your risk all round. The same is true if a homocysteine test is argued to predict low B vitamin status only. Once again, this leads to high H scores, increased arterial disease, and Alzheimer's.

■ Does high homocysteine cause Alzheimer's?

So, the big question is which comes first – a high H score or Alzheimer's? Doctors from the Department of Neurology at the Boston University School of Medicine wanted to determine whether homocysteine precedes mental decline, or occurs as a result of dementia-related B vitamin deficiencies.

Their study looked at 1,092 people who had an average age of 76 and did not have dementia (a brain disorder associated with loss of memory, concentration, and judgement, sometimes accompanied by emotional disturbance and personality changes). These people had already taken part in another study measuring their homocysteine levels eight years earlier. Researchers again measured their homocysteine levels, and then kept track of their mental health over the next eight years. During that time, 111 developed dementia, 83 of whom were diagnosed with Alzheimer's. The findings revealed that the higher the homocysteine levels preceding any symptoms of mental decline, the greater the risk of later developing dementia. In those with a homocysteine score of more than 14 units, the risk of Alzheimer's almost doubled.[40] They concluded that 'an increased homocysteine level is a strong, independent risk factor for the development of dementia and Alzheimer's disease'.

More recently, evidence has emerged that even before there is evidence of declining mental function in so-called 'healthy' elderly individuals, high levels of homocysteine also predict physical degeneration in certain parts of the brain.[41]

A research group at the Baylor University Metabolic Disease Center in Dallas, Texas, led by Dr Teodoro Bottiglieri – one of the world's leading experts in the connection between folate and mental illness – suggested that low levels of folate (leading to raised homocysteine) may cause brain damage that triggers dementia and Alzheimer's. Their research has found that a third of those with both dementia and high H scores (above 14 units) are deficient in folate.[42]

Alzheimer's sufferers also have less SAMe in their brains, as well as higher levels of homocysteine in their blood. As we saw in Part 1, SAMe, derived from the methylation of homocysteine, is the brain's single most important methyl donor, helping produce and activate all sorts of neurotransmitters, including the memory enhancer acetylcholine – and levels of this decline in Alzheimer's.[43]

These findings are being echoed by research around the world. In Scotland, researchers have found that reduced mental performance in old age is strongly associated with high levels of homocysteine and low levels of vitamins B12 and folate. They studied people who had taken part in the Scottish Mental Surveys of 1932 and 1947, which surveyed childhood intelligence. And they found that while homocysteine was higher and mental performance weaker in the older group, the most mentally agile of either group had the highest levels of B vitamins and the lowest levels of homocysteine. In the older group, high levels of homocysteine accounted for a 7 to 8 per cent decline in mental performance.[44]

Whichever way you cut it, the accumulating evidence is pointing to a consistent pattern. The higher your H score and the lower your B vitamin status, the greater your chances of declining memory, poor concentration and judgement, lowered mood, physical degeneration and poor circulation to the brain. Exactly how high homocysteine – and the inevitable B vitamin and SAMe deficiencies that always accompany high homocysteine – might contribute to the kind of brain damage seen in Alzheimer's is yet to be discovered. However, if simply following the H Factor diet and supplement programme can prevent this, what are we waiting for?

14

Thyroid problems linked to homocysteine

THE SECOND MOST commonly prescribed drug in the US is the synthetic thyroid hormone levothyroxine. This is because an estimated 27 million people in the US have an underactive thyroid, along with 1.4 million in Britain. Thyroid problems are also very much on the increase – recent research suggests that well over 5 per cent of the population probably have underactive thyroids,[45] yet only half are diagnosed. An underactive thyroid means a drop in the production of the hormone thyroxine, which raises the activity levels of cells throughout the body.

The classic symptoms and signs of an underactive thyroid include:

- Chronic fatigue or lack of energy

- Weight gain and an inability to lose weight on a low-calorie diet

- Feeling cold, often when others feel comfortable

- Dry skin and dry, thin, prematurely greying hair

- Yellowish tint to palms, soles and roof of mouth

- Infertility

- 'Dirty looking' elbows and knees

- High cholesterol

- Increased risk of heart disease

- Heavy menstrual periods

- Constipation

- Slowed thinking

- Short fifth finger, especially in women.

Thyroid problems can occur for two main reasons. The first is underactivity – or hypothyroidism – as an indication of exhaustion and ageing. This is very much linked with stress, blood sugar problems, and/or a poor diet associated with iodine, zinc and/or selenium deficiencies. The second is an autoimmune reaction where the body wrongly attacks its own thyroid cells, leading to poor production of thyroxine (called Hashimoto's thyroiditis). Autoimmune reactions can also go the other way, causing thyroid cells to multiply or swell, much like cancer cells, leading to the release of too much thyroxine and an overactive thyroid – or hyperthyroidism (called Grave's disease). There is growing evidence that many people with an autoimmune thyroid disease may actually be experiencing an allergic reaction to wheat and other gluten grains.

Autoimmune thyroid problems are detected by measuring the presence of anti-thyroid antibodies in the blood. Recent research is finding that 11 per cent of all people have these antibodies and are therefore especially at risk of developing thyroid problems.

■ The homocysteine link

Whatever the cause, there is a direct link between low levels of thyroid hormone and high levels of homocysteine. Endocrinologists at the Cleveland Clinic Foundation and the University of California, Davis, have found that hypothyroidism is associated with high homocysteine levels. Following drug treatment to normalise the thyroid, homocysteine levels also return to normal.[46]

When levels of thyroxine are low, the pituitary gland in the centre of the head releases more thyroid stimulating hormone (TSH) to stimulate the thyroid gland to produce and release more thyroxine. Once again, there is a clear association between high levels of homocysteine, low levels of thyroxine, and in response to low levels of thyroxine, high TSH. When homocysteine returns to normal, so does thyroxine and TSH, and vice versa.[47]

Research all over the world – from the US and UK to the Czech Republic, Denmark to France – is finding this very strong link between homocysteine and thyroid problems. Danish scientists have found that the higher your homocysteine, the lower your thyroxine level is likely to be, and the greater the likelihood that you will experience symptoms and signs of underactive thyroid. So they now recommend that anyone with a high H score be routinely checked for thyroid problems.[48]

Hypothyroidism is also a risk factor for atherosclerosis, in which fatty deposits clog up the arteries and increase blood pressure. Researchers at the Jean Mayer US Department of Agriculture Human Nutrition Research Center on Aging compared subjects with hypothyroidism, high homocysteine and cholesterol levels.[49] They found that people with hypothyroidism were 4.6 per cent more likely to have high H scores (above 12 units) and 8 per cent more likely to have high cholesterol (above 6.2 mmol/l). The research concluded that high levels of homocysteine and cholesterol 'could help to explain the increased risk for arteriosclerotic coronary artery disease in hypothyroidism'.

While thyroid hormones can reverse the symptoms in patients with hypothyroidism, research has found that thyroid hormones are not nearly as successful at lowering high homocysteine levels as the right combination of B vitamins and other supplements.[50] Not surprisingly, hypothyroid sufferers also have lower levels of folate, indicating a greater need for this vitamin. All these findings are indicating that thyroid problems may be better treated by lowering homocysteine with the H Factor programme, combined with appropriate thyroid medication.

15

Depression and schizophrenia – a methylation problem?

ACCORDING TO THE World Health Organization, mental health problems are fast becoming the number one health issue this century. Ten per cent of people worldwide are suffering mental problems at any one time, and 25 per cent or 1.5 billion humans will suffer from them at some point in their lives.[51] Schizophrenia and depression are increasing the world over. Depression is ten times more common today than it was in the 1950s, despite improvements in social and economic conditions. It is also the primary cause of suicide, claiming 3,000 lives a year in England and 30,000 in America, and is now the second most common cause of death in young people aged 15 to 24.

One of the greatest shortcomings of conventional psychiatry is the failure to check for basic biochemical and nutritional imbalances, such as blood sugar problems, food allergies, gluten sensitivity, and vitamin, mineral and essential fatty acid deficiencies, in people with mental health problems. More than any other organ in the body, the brain is totally dependent on a second-by-second supply of nutrients. It is constructed out of nutrients, communicates via nutrient-dependent neurotransmitters, is protected by nutrients and fuelled by nutrients. When you consider that the brain is entirely made from molecules derived from food, air and water, and that simple molecules like alcohol can fundamentally affect the brain, isn't it exceedingly likely that changes in diet and the environment affect our mental health?

If you are suffering from some of the following symptoms, the chances are you are not getting all the nutrients your brain needs to keep you happy and optimistic:

- Feelings of worthlessness or guilt

- Poor concentration

- Forgetfulness

- Poor memory

- Loss of energy, and fatigue

- Thoughts of suicide or preoccupation with death

- Loss or increase of appetite and weight

- Disturbed sleeping pattern

- Slowing down (both physically and mentally)

- Irritability, agitation (restlessness or anxiety)

- Loss of interest in previously enjoyable activities or people

- Addictive cravings and behaviour.

Top of the 'suspect' list when it comes to depression and schizophrenia are the methyl-moving, homocysteine-lowering B6, B12 and folate nutrients.

■ Depression is often a folate deficiency

If you're depressed, you may well be low in folate. In a study of 213 depressed patients at the Depression and Clinical Research Program at Boston's Massachusetts General Hospital, people with lower folate levels had more 'melancholic' depression and were less likely to improve when given anti-depressant drugs.[52] Very depressed people, and also those diagnosed with schizophrenia, are also often deficient in folate. A survey of such patients at Kings College Hospital's psychiatry department in London found that one in three had borderline or definite folate deficiency. These patients then took folate for six months in addition to their standard drug treatment. The longer they took the folate, the better they felt.[53]

Folate, B12 and vitamin B6 help 'tune up' the brain, and by supporting homocysteine metabolism they increase production of SAMe,[54] another intelligent nutrient that helps the brain work better. Research has shown that SAMe is itself an effective natural anti-depressant, and well worth supplementing at 200 to 1200mg a day. (You need to pick SAMe supplements with care, however; see Part 5, page 216 for more about this.) All of these B vitamins not only help the brain make more serotonin (which makes us feel happier, sleep better, experience less physical pain, and become less agitated and less apt to crave carbohydrates), but they are masters of methylation, helping to keep the brain's delicate chemistry in balance by moving methyl groups to make new chemicals as required.[55]

■ Methylation and mood

You might have wondered why supplementing a large amount of nutrients such as folate – way above what you can eat in a well-balanced diet – seems to be so effective. The answer is that we are biochemically unique and there is growing evidence that many people, perhaps those prone to severe depression or schizophrenia, have high homocysteine levels, and low B vitamin and SAMe status and hence are not methylating properly. Methylation is a chemical process that is absolutely vital for the brain, as well as the body, helping to turn one neurotransmitter into another (see Chapter 36). Nutrients that can donate or receive methyl groups help the brain to function much better.

Homocysteine is strongly linked to depression

Your homocysteine level is the best indicator of whether you are getting enough of these critical B vitamins to have a high methyl IQ (meaning, to be able to methylate effectively). It therefore comes as no surprise to find that people with severe depression and schizophrenia often have high H scores. In one study, more than half (52 per cent) the women participating, who had severe depression, were found to have elevated homocysteine and low levels of folate.[56] In another study, a significant proportion of patients with schizophrenia had increased homocysteine levels despite no evidence of dietary deficiency in folate or B12.[57] (Very important to remember: your homocysteine level is a better method of detecting B vitamin deficiencies – particularly deficiencies of folate and B12 – than conventional blood tests.)

Homocysteine levels are particularly high in young men diagnosed with schizophrenia. When comparing 193 men and women with schizophrenia with 762 who did not have schizophrenia, researchers in the US found that average homocysteine levels were a very high 16.3 units for schizophrenics compared to 10.6 units in normal subjects.[58] But the difference between the groups was almost entirely attributable to the homocysteine levels of young male patients with schizophrenia.

Men experience most mental health problems, including depression and schizophrenia, between the ages of 13 and 18, when these conditions can surface or get worse. This is the time when the body grows faster than the brain, demanding a greater proportion of nutrients, at the same age when nutrient-depleting junk food, cigarettes, alcohol and illicit drugs become more popular. Any young man on the edge of nutritional insufficiency can easily be tipped over the edge. The same does not seem to apply to young women, whose growth rate is much more consistent.

■ Genes, homocysteine and mental illness

Not everyone is born equal, as far as methylation is concerned. As you'll see in Part 4, around one in ten of us inherit a defective gene that means a methylating enzyme called MTHFR doesn't work so well. This creates a logjam in the 'dance' of methyl groups, an increase in homocysteine, and a corresponding increased need for B vitamins (see Figure 3 on page 25).

The case of a 27-year-old woman diagnosed with schizophrenia, who had a high H score, illustrates this point well. Her doctors at the Department of Clinical Neuroscience in Molndal Hospital in Goteburg, Sweden, found that giving her B12 injections improved her condition. When they stopped, her schizophrenia deteriorated. However, the beneficial effects of prolonged weekly B12 treatment appeared to diminish as time went on, suggesting that the abnormality was not wholly B12-dependent. They found that she also had the defective MTHFR enzyme, which explained her increased levels of homocysteine.[59] This case highlights the importance of measuring homocysteine in anyone diagnosed with depression or schizophrenia.

While folate deficiency alone can make you crazy or depressed, the combination of folate deficiency and a fault in the MTHFR gene is more than likely to trigger mental illness. To compensate for this, much higher levels of folate than normal are needed. According to research from Columbia University's Department of Psychiatry in New York, this applies to people with schizophrenia. They found increased levels of homo-cysteine, despite no apparent lack of folate.[60] The same is true for vitamin B12. Many people with mental illness need more than a normal amount of vitamin B12 despite no obvious signs of advanced deficiency such as anaemia.[61]

A far better indicator of increased personal or individualised need for these B vitamins is your H score. This is why it is so essential to test and periodically retest your homocysteine level, rather than just relying on eating greens or supplementing basic levels of B vitamins.

16

Parkinson's disease and homocysteine

PARKINSON'S IS A frustrating and upsetting disease, and it isn't confined to old age. It affects around 120,000 people in Britain and a million in the US, from teenagers to the elderly. The signs – hand tremor, rigidity, emotionless face, difficulty in changing directions when walking and unsteadiness – are caused in large part by a lack of dopamine, an essential neurotransmitter or chemical messenger, in the brain and digestive tract.

Why this deficiency occurs is still a bit of a mystery. Part of the puzzle is enzyme deficiencies that turn dietary amino acids into dopamine. Optimum nutrition, including large amounts of certain B vitamins, NADH (a coenzyme made in part from the B vitamin niacin), zinc and magnesium, often helps. (Note that most of these have been identified as homocysteine-lowering nutrients.) According to researchers at Boston University, having a high H score is also a strong, independent risk factor for developing Parkinson's disease.[62] There is also growing evidence that glutathione deficiency in the brain of Parkinson's patients plays a key role. We therefore recommend that every person diagnosed with Parkinson's be tested for homocysteine, and if that measures above 9 units, be immediately placed on the H Factor programme until their homocysteine has optimised below 6. This alone is likely to provide considerable relief.

■ Homocysteine is a neurotoxin

The main cause of Parkinson's is thought to be damage to the dopamine-producing brain nerve cells, or neurons. Homocysteine, which is itself toxic to nerves, may have a role to play here, too.

Researchers at the Laboratory of Neurosciences in Bethesda, Maryland, set out to discover the role that folate and homocysteine play in this damage. They found that when mice were deprived of folate, predictably their homocysteine levels went up, and they became especially vulnerable to a neurotoxin that destroys these dopamine-producing neurons. Homocysteine was not only found to exacerbate this toxin, but also to cause direct damage to the neurons itself. If, however, the mice were fed additional amounts of folate, they were able to repair the damage in dopamine-producing neurons and counteract the adverse effects of homocysteine.[63]

While it is early days in researching the link between homocysteine and Parkinson's disease, the strong implication here is that maintaining an optimal H score, and hence folate, as well as B12, B6 and B2, and nutritional status, will prevent the development of Parkinson's, protect against brain damage, and improve the symptoms.

17

Pregnancy problems and infertility

BABIES DEMAND THE BEST: no time is more important for optimum nutrition than just before conception, throughout pregnancy, and the first four to six months of exclusive breastfeeding. First, any deficiency or imbalance affects two people, not just one. Secondly, during pregnancy and breastfeeding, much higher intakes of nutrients are needed to optimally nourish a developing foetus and infant. So women on the brink of inadequate nutrition often fall over the edge, and become infertile, have repeated miscarriages, premature births or other pregnancy problems, or sometimes bear children with mental or physical defects.

So strong is the link between these problems and homocysteine that we strongly recommend that ANY woman who either intends to get pregnant or has just become pregnant have a homocysteine test as soon as possible. If your H score is high, you know that you are insufficiently nourished in one or more key B vitamins, zinc and/or magnesium, are methylating poorly and are deficient in SAMe (an important methyl donor), glutathione (a powerful antioxidant and detoxifying agent), nitric oxide (another potent antioxidant, neurotransmitter and artery wall relaxant) and other critical biochemicals. In an ideal world, a woman should not get pregnant until her H score is below 6 because this minimises the risk of problems in pregnancy. Let's examine the evidence.

■ High homocysteine increases the risk of pregnancy problems

As part of a large-scale study in Norway called the Hordaland Homocysteine Study, investigators measured homocysteine in over 5,800 pre-menopausal Norwegian women between the ages of 40 and 42. They compared homocysteine levels with the women's previous pregnancy outcomes as recorded by a medical birth registry.[64]

Women with higher homocysteine levels were 32 to 101 per cent more likely to have experienced pregnancy complications such as pre-eclampsia, premature birth and low birth-weight babies. Birth defects including neural tube defects (such as spina bifida) and club foot also occurred more frequently in babies whose mothers had higher levels of homocysteine, a finding that has been confirmed by other research groups.[65]

A separate Dutch study found a similar result and concluded that high homocysteine and/or low folate (a nutrient particularly key in pregnancy) can indicate risk factors for repeated early pregnancy loss. Among women who experienced repeated miscarriages, those with higher homocysteine levels had roughly a three- to fourfold greater risk of suffering further miscarriages.[66] Women who have had early miscarriages are also more likely to have babies born with defects, both of which are more common if your H score is higher.[67]

The good news is that knowing your H score doesn't just predict whether you're at risk. It also provides the answer to reducing that risk – namely, the H Factor diet and supplement programme. To repeat: we strongly believe every woman thinking about becoming pregnant and every woman just informed that she's pregnant should be tested for homocysteine levels and advised accordingly.

Especially at risk are the 10 per cent of pre-menopausal women who have a common genetic defect impairing their ability to produce an enzyme (MTHFR) that converts potentially toxic homocysteine back into methionine and then into SAMe. These women therefore have a tendency towards high homocysteine levels, requiring much larger intakes of homocysteine-lowering nutrients to achieve an optimal level. The same applies to the baby. Embryos that are miscarried are much more likely to have inherited the faulty MTHFR gene that leads to high levels of homocysteine.[68]

Time and time again, research shows that diet alone isn't enough to address these pregnancy problems. To achieve optimal health and a healthy pregnancy, supplementation both before and during pregnancy is essential.

Interestingly, in the Dutch study mentioned earlier, women with repeated early miscarriages were nearly twice as likely to have this MTHFR genetic defect, compared to women without miscarriages.

The medical profession is starting to wake up to the need to monitor homocysteine before and during pregnancy. A recent editorial in the *American Journal of Clinical Nutrition* by Dr Mary Frances Piccianno of the National Institutes of Health in Maryland stated that any factors that raised homocysteine were:

> probable risk factors for [problem pregnancies including] placenta-mediated diseases, such as pre-eclampsia, spontaneous abortion and placental abruption. The need for research is evident when considering that in 1996 nearly 30,000 US infants died before their first birthday, and birth defects, pre-term birth, low birth weight, and maternal and placental complications were among the leading causes.

■ Defining optimum nutrition in pregnancy

As we've said, folate is key to a healthy pregnancy, and supplementing with at least 400mcg a day is well proven to minimise the risk of birth defects. But folate is by no means the only essential nutrient to supplement for pregnant women.

The importance of folate in pregnancy first came on the map back in 1970, when Dr Richard Smithells reported a link between insufficient folate and spina bifida, a congenital defect in which the spinal column is imperfectly closed at birth, often resulting in permanently disabling neurological disorders. It took 20 years for Dr Smithells's work to be taken seriously and then, in 1992, for the first time ever, the US and UK governments recommended a nutritional supplement, saying that pregnant women should supplement 400mcg of folate daily, an amount not easily achieved by diet alone. This is the first crack ever to appear in the false medical belief that 'as long as you eat a well-balanced diet, you get all the nutrients you need'.

It is essential to supplement daily with folate prior to and throughout pregnancy, possibly in larger amounts depending on your H score. Folate both reduces the risk of birth defects and miscarriages[69] and prevents premature birth and low birth weight.[70]

So does another homocysteine-lowering nutrient, vitamin B12. In a study of 110 women who had a history of repeated early miscarriages, vitamin B12 status was found to be lower and homocysteine levels higher than a control group of women with one or more children and no miscarriage history. Giving vitamin B12 supplements to these women led to four normal pregnancies in the five who became pregnant again.[71] In addition to insufficient folate, B12 is also often low in women who have had babies born with neural tube defects, such as spina bifida.[72]

Another key homocysteine-lowering vitamin is B6. In a study

of Chinese women aged between 21 and 43 years who'd lost a baby, researchers measured vitamin B6 status as well as homocysteine and folate levels. Compared to a control group (who'd had a full-term pregnancy ending with a live birth), levels of B6 were lower – with the risk of miscarriage being greater the lower B6 levels were. Miscarriage risk was four times more likely when folate was also low along with B6.[73]

Other essential nutrients during pregnancy include zinc and omega-3 fats (found in oily fish, flaxseed and nuts), both thought to be important in optimising homocysteine and methylation, which as we've seen is the chemical process that transforms one biochemical into another. Zinc is especially important to supplement since the optimal intake of zinc is 20mg during pregnancy, which is way above the average daily intake of 7.6mg.

To repeat, zinc, vitamin B6, B12 and folate are among the most important nutrients for lowering homocysteine (vitamin B2, TMG and probably magnesium are the others), and your H score is really the best way of knowing whether or not you are getting enough of these nutrients. So a person's need is very simply whatever keeps your H score below 6 units.

▪ Homocysteine damages the placenta

Having a high homocysteine level not only indicates whether you are optimally nourished with these top 'methyl movers' but it also damages a key organ of pregnancy, the placenta. As a result, other essential nutrients, such as the essential omega-3 fat called DHA, can't be well enough supplied from mother to baby during pregnancy. Babies born with birth defects consistently have lower DHA levels and high homocysteine levels.[74] What is of more concern is that children are being born with high homocysteine levels, which stay high even if breastfeeding, if their mothers have high homocysteine levels (one exception being Down's syndrome, where the mother has high homocysteine but the Down's infant has low levels) – although both of these conditions are correctable with the right nutrients.[75]

▪ Homocysteine and Down's syndrome

People tend to think that 'genetic' diseases such as Down's syndrome are both unpreventable and untreatable. These beliefs are untrue. Genes become damaged for a number of reasons, and often because of high levels of homocysteine, faulty methylation, and low levels of glutathione, which are vital for protecting and repairing DNA. Researchers at the Department of Paediatrics at United Arab Emirates University believe that low folate status combined with a fault in the MTHFR gene (which ordinarily produces an enzyme needed to break down homocysteine) and high homocysteine in the mother can result in Down's syndrome. They cite the case of a mother with this profile (an extremely high homocysteine score of 25 units) who gave birth to a child with Down's syndrome and neural tube defects.[76] Much more research is needed to confirm whether the same outcome occurs in similar cases.

While there is growing evidence that mothers with a high H score and hence low B vitamin status may be more at risk of a Down's syndrome child, it's a paradox that Down's syndrome children themselves have very low homocysteine levels. This is because the section of genes that is damaged in infants and children with Down's syndrome include one of the key methylation genes. This enzyme becomes overactive and consequently these children have less homocysteine. But they also have less of the amino acid methionine, which is used to make SAMe (our body's main methyl donor – vital for lots of other essential functions). So while the low homocysteine sounds like good news, in the case of Down's syndrome it's very bad news, because Down's children are unable to make enough methionine or its end product, SAMe, the human body's primary methyl donor. Consequently, these children are SAMe and methionine deficient, and need to get much more folate and methyl donors from diet and supplements.

One investigation showed that it is possible to correct SAMe-related problems in Down's cells in a test tube by using variants of folate and vitamin B12. [77] In addition, many parents of Down's children have reported improvements in their son or daughter after supplementation with therapeutic doses of methyl donors such as TMG and the methyl movers folate and B12. We recommend that they also consider supplementing with therapeutic doses of SAMe.

▪ Homocysteine and infant epilepsy

Epilepsy may also be linked to high homocysteine. A small study of ten boys and girls who suffered from epilepsy found that those who had experienced seizures in their first year of life all had high levels of homocysteine. When they were given folate and B12 supplements, however, neither their incidence of seizures nor their homocysteine were reduced. [78] We strongly suspect they were not given enough of the right combination of nutrients, including B6, B2 and magnesium, all well known anti-epileptic nutrients. Such a strategy is a foolproof way of lowering homocysteine and, had this been achieved, a reduction in epilepsy symptoms may have followed.

▪ Maximise your fertility by lowering your homocysteine

There's something about lowering your homocysteine level that just may increase your fertility, whether you're a man or a woman. There are many reasons for infertility in either sex. These include lack of nutrients such as zinc, essential fats and B vitamins, untreated coeliac disease, an underactive thyroid, lack of progesterone, ovarian diseases including polycystic ovary syndrome, and so on. We think high levels of homocysteine (often associated with zinc, essential fats and B vitamin deficiencies, as well as coeliac disease and underactive thyroid) should be added to the list, even though the evidence is not yet extensive.

In men, a high level of homocysteine is strongly associated with low sperm motility. Motility is the 'swimming power' of sperm, which determines whether they can make it to the egg and penetrate it, or run out of puff. In one study, a high level of homocysteine was associated with 57 per cent less motility![79] This may be because methylation, the chemical process that homocysteine reflects, is absolutely vital for healthy sperm production.

One of the common causes for infertility in women is polycystic ovary syndrome (PCOS). Women with PCOS have a higher cardiovascular disease risk profile, including more chance of developing atherosclerosis. This is usually attributed to higher blood sugar and fat levels, as PCOS sufferers can have difficulty controlling insulin production. However, research from Italy has found that women with PCOS are much more likely to have high homocysteine levels.[80] So swapping refined carbohydrates and sugar for whole (non-gluten) grains and more fruit and vegetables, reducing intake of saturated fats from meat and dairy foods while increasing essential fats from oily fish, flaxseed and nuts can help to reverse symptoms and restore fertility. As too can lowering a high H score by ensuring an adequate supply of the right nutrients (see Part 5).

Perhaps an even more common cause of infertility in women is undiagnosed sensitivity to wheat and other gluten grains, called coeliac disease. By continuing to eat gluten grains, the sensitive women create a digestive tract that can't absorb nutrients well. Which nutrients? You guessed it: all the well-established homocysteine-lowering folate, B12, B6, B2 and zinc – and so deficiencies occur, resulting in high homocysteine levels. Eliminating gluten from the diet is the treatment of choice with these gluten-sensitive women. Supplementation with the missing nutrients and monitoring your H score is also encouraged. This strongly suggests that infertile women should follow the H Factor programme.

18

Chronic pain – from arthritis to migraines

PAIN IS THE BODY'S alarm system, telling you it's gone seriously out of balance. Usually, pain goes hand in hand with inflammation – aching and discomfort, redness, soreness or swelling. However, inflammation can go on unnoticed. Alzheimer's, Parkinson's, heart disease, diabetes, depression and cancer, for example, all involve inflammation to a greater or lesser degree.

When the body has lost its ability to methylate properly, which is what having a high H score is all about, pain and inflammation are just around the corner.

Let's examine the evidence for the most common cause of pain – arthritis. Although there are over 100 different kinds of arthritis, there are two very prevalent varieties of this disease. The most common by far is osteoarthritis, but the most debilitating and destructive is rheumatoid arthritis. In this autoimmune disease, the body's immune system, instead of protecting it from disease, reacts in a way that damages the joints and other tissues of the body.

About 80 per cent of people over the age of 50 show

osteoarthritis-like joint damage, although only a quarter of them experience pain. By the age of 60, more than 90 per cent of people show evidence on X-rays of arthritis-like joint damage. While osteoarthritis occurs later in life, painful and stiff knee problems – often diagnosed as chondromalacia, an abnormal softening or degeneration of joint cartilage, especially of the knee – occur frequently in people under 40.

Under the age of 45, osteoarthritis is more common in men. Over the age of 45, it's more common in women. It starts as stiffness, usually of the weight-bearing joints such as the knees, hips and back, and then progresses to pain on movement. The joints then become increasingly swollen and inflexible.

Rheumatoid arthritis can develop in people much earlier – some as young as 25, although the peak age is 30 to 50. There are an estimated half a million people in Britain and 3 million in America who suffer with this type of arthritis, and around 80 per cent of them are women.

▪ Rheumatoid arthritics have high homocysteine levels

Since rheumatoid arthritis is a 'systemic' disease, where the whole body's chemistry is out of balance and many tissues and organs other than the joints are affected, one would suspect that homocysteine plays a leading role in the disease. And it does. Research from the Department of Biochemistry at the University Hospital in Madrid, Spain, examined the H scores of women with rheumatoid arthritis versus those without.[81] There was a massive difference. The average H score for those with rheumatoid arthritis was a sky-high 17.3, compared to 7.6 for those without!

Other research groups have found similar differences, especially among rheumatoid arthritis sufferers with a history of thrombosis,[82] or abnormal clotting of blood. Homocysteine has also been implicated in ankylosing spondylitis, an inflammatory, crippling arthritic disease of the spine.[83]

Homocysteine is thought to damage joints and other tissue directly, as do oxidants produced by the body. Remember, the presence of high homocysteine levels means there's less glutathione available. This incredibly powerful antioxidant minimises the damage wreaked by oxidants. When it's deficient – as with high levels of homocysteine – the oxidants take over, accelerating the destruction of joint tissues. This can speed the development of both rheumatoid arthritis and osteoarthritis. The low SAMe levels that hit when homocysteine is high are also significant here, as SAMe has proven anti-osteoarthritic benefits.

All of this suggests that reducing homocysteine with the H Factor programme may well help those with arthritis considerably. Yet, disappointingly, very little research has yet been done to test the homocysteine theory of arthritis. An exception occurred in 1994 when forward-thinking researchers from the highly esteemed American College of Nutrition in Clearwater, Florida, gave B12 and folate supplements to 26 people who had been suffering from osteoarthritis of the hands for an average of more than five years, and had been taking non-steroidal anti-inflammatory, pain-relieving drugs (NSAIDs). The results showed that people with arthritis who took the vitamins in place of the NSAIDs had less tenderness in their hand joints, and similar improvement in their ability to grip objects, compared with those just taking NSAIDs, but none of the notorious side-effects seen with the use of these drugs.[84] NSAIDs can cause serious reactions, including premature death from kidney failure, ulcers and bleeding in the digestive tract, and they cost considerably more than B12 and folate supplements.

B6, another homocysteine-lowering vitamin, may also help arthritis sufferers. Back in the 1950s, an insightful physician from Mount Pleasant in Texas, Dr John Ellis, found that giving B6 in higher daily doses of 50 mg helped to control pain and restore joint mobility to his arthritic patients.[85] Vitamin B6 shrinks inflamed membranes that line the weight-bearing surfaces of the joints, perhaps by helping decrease homocysteine and increase SAMe and glutathione, both proven anti-inflammatory agents. B6 also helps to regulate production of the prostaglandins, the body's own anti-inflammatory agents. We predict that following the H Factor programme will produce even more spectacular results.

■ Osteoporosis, oestrogen and homocysteine

Although not an inflammatory disease as such, osteoporosis is very much linked to high homocysteine. A decrease in bone mass density resulting in osteoporosis is a common symptom of the rare genetic disease homocystinuria, which is associated with very high homocysteine levels. Recent research from Japan has found that women with the greatest post-menopausal bone mass loss are much more likely to have a defect in the gene (MTHFR) that produces an enzyme that detoxifies homocysteine, resulting in a higher than normal H score.[86] These women may be particularly at risk of osteoporosis unless they follow a diet and supplement plan to lower their H score.

A low level of oestrogen, which is very common in post-menopausal women, also appears to raise homocysteine and increase osteoporosis risk. Theoretically, increasing oestrogen could help lower homocysteine.[87] This has been shown in some preliminary studies,[88] but not in others. Until more is understood about how homocysteine and oestrogen are related, we are reluctant to recommend oestrogen HRT as a means to lower homocysteine for two reasons. One is that it is less effective than the nutritional strategy making up the H Factor programme; the other is that it carries an increased risk of breast and uterine cancer. 'Natural' progesterone HRT (not to be confused with synthetic progestins used in most HRT preparations) does not have these associated risks. But no one has yet investigated whether this lowers H scores.

■ Fibromyalgia, chronic fatigue and homocysteine

Another common painful condition is fibromyalgia. This refers more to muscle pain and tender spots, than to joint pain. Any person with fibromyalgia, sometimes also diagnosed as polymyalgia, will tell you that the pain is just the tip of the iceberg. It is often accompanied by chronic, disabling fatigue; many very tender spots on the neck, shoulders, back and hips; constant aches; general stiffness; sleep disturbances; depression; allergic rhinitis – and sometimes even cardiovascular problems. As such, it is similar in a number of ways to chronic fatigue syndrome. But what, if anything, has this got to do with homocysteine?

A team of researchers headed by Dr Bjorn Regland at the Institute of Clinical Neuroscience at Sweden's Goteborg University decided to find out. They ran a battery of tests on fibromyalgia sufferers, including homocysteine. By far the most significant finding was that every single patient with fibromyalgia had high levels of homocysteine. They also found a direct correlation between their B12 status and the severity of their reported symptoms.[89] Fibromyalgia sufferers should be routinely tested for homocysteine, and if high, immediately started on our H Factor programme.

■ Are migraines an inherited high homocysteine condition?

Anyone who has experienced migraines (and there are over 30 million migraine sufferers in the US alone, 10 million of whom see doctors regularly for relief) knows how excruciatingly painful they can be. A migraine is a severe, recurring headache triggered by both constriction and dilation of the blood vessels in the head. Homocysteine, you may recall, has a profound effect on blood vessels, even triggering stroke and cardiovascular disorders. So Dr H. Kowa, from the Institute of Neurological Sciences Faculty of Medicine at Japan's Tottori University, wondered whether homocysteine might have anything to do with migraines.

Dr Kowa recruited 74 patients who had frequent migraine headaches and 261 normal, healthy controls. After testing them for the MTHFR gene mutation, which indicates a tendency to overproduce homocysteine, he found that compared to controls more than twice as many of the migraine sufferers had the mutation. And sufferers who experienced 'aura' symptoms before a migraine – such as blurred vision, bright spots in their field of vision, muddled or confused thinking, extreme exhaustion, anxiety, numbness or a tingling sensation in one side of the body – were four times more likely to have the mutation and high levels of homocysteine.[90]

This study suggests that the tendency to migraines might be inherited in many people due to the MTHFR gene mutation, and that high homocysteine levels might also be involved. That has yet to be researched, but if so it would suggest that a homocysteine-lowering diet and supplement programme might prove enormously helpful for migraine sufferers.

19

Digestive disorders, stomach ulcers and coeliac disease

IT MIGHT SEEM A BIT ODD, but think of the digestive tract as your 'inner' skin. And there's a lot of it: if the inside lining of the small intestine were laid out flat, it would cover a surface area larger than a tennis court. It's through this surface that all essential nutrients from food are absorbed. Any problems with absorption almost invariably lead to lower levels of vitamins, including B6, B12, folate, B2 and zinc, and, consequently, higher levels of homocysteine.

■ Gluten, coeliac disease and homocysteine

Coeliac disease, also known as gluten gastroenteropathy, is an inherited extreme sensitivity to gluten, a protein which is found in wheat, rye, barley and oats. The only proven therapy is strict lifelong elimination of all gluten from the diet. The official opinion in the US, until very recently, is that coeliac disease is very rare, occurring in fewer than one in 5,000 people. But this view is out of date. In fact, it's very common, occurring in about one in every hundred apparently healthy people and much more commonly in people with chronic illnesses such as migraines, Type I diabetes, autoimmune thyroid disease, certain cancers, arthritis and osteoporosis. This makes coeliac disease over twice as common as cystic fibrosis, ulcerative colitis and Crohn's disease combined! New research indicates that it may occur in as many as 200 different medical conditions, but very frequently goes undiagnosed (see Dr Braly's recent book, *Dangerous Grains*, for more details).

Some of the more important symptoms and conditions associated with coeliacs can include:

- Upper respiratory tract problems like sinusitis and otitis media (fluid build-up and/or infection of the middle ear)

- Malabsorption of essential nutrients, even when supplemented orally, causing very high homocysteine, fatigue, anaemia, certain cancers, autoimmune diseases, osteoporosis and weight loss

- Diarrhoea, constipation, bloating, distention, Crohn's disease, ulcerative colitis, diverticulitis, *Helicobacter pylori*-caused stomach ulcers

- Depression, chronic fatigue syndrome, and behavioural problems in children such as ADHD and social conduct disorders.

In people with coeliac disease there is always damage to the small intestine, leading to poor absorption of B vitamins and zinc and therefore high homocysteine levels. Since coeliac disease so often goes undiagnosed (for every one diagnosed, 39 go undiagnosed and untreated), we believe that people with very high H scores that don't fall within the safe zone, even after following the H

Factor programme, should be routinely checked to see whether they have the disease. If you test positive for coeliac disease and gluten sensitivity, lifelong elimination of all dietary gluten is the treatment of choice.

Researchers from Norway have found that a high level of homocysteine, probably reflecting deficiency in B vitamins, is very common among those with other digestive disorders[91] in addition to coeliac disease. They had 24 patients with extremely high H scores, above 40 units, who had gastrointestinal diseases, including coeliac disease, Crohn's disease, ulcerative colitis, gastritis and *Helicobacter pylori* infection, which is the most common cause of stomach ulcers. They wondered whether these people might have the genetic mutation (MTHFR) that leads to higher levels of homocysteine. Nineteen out of the 24 did, which means that they effectively needed much more B6, B12, folate, B2 and zinc than others, and yet were less able to absorb these due to their digestive problems. This is a vicious circle, since homocysteine itself may further damage the digestive tract. In this case, to alleviate the symptoms of chronic digestive disorders and reduce a high H score, the H Factor programme may not be enough on its own. We would recommend a tailor-made treatment programme from a nutrition therapist.

▓ Stomach ulcers, *Helicobacter pylori* and homocysteine

A common cause of stomach ulcers is infection with a bacterium called *Helicobacter pylori*. People with stomach ulcers and gastritis are frequently infected with *H. pylori*, and this often leads to poor absorption of B vitamins, especially B12. This is because B12 has to combine with 'intrinsic factor', a stomach chemical, to become absorbable further down the digestive tract. The end result of deficiency in B12 and other B vitamins is high homocysteine, and an increased risk of, among many others, diseases of the heart and brain arteries.[92]

Some researchers have started to trace this line of events backwards, wondering whether those with heart attacks or strokes might have undetected *H. pylori* infection leading to poor absorption of B vitamins, thus raising homocysteine. Antonio Piertroiusti, professor of internal medicine at Tor Vergata University in Rome, and his team, tested for *H. pylori* infection in the blood of patients who had had strokes caused by severely narrowed arteries. Of these stroke victims, 42 per cent showed evidence of *H. pylori* infection, compared to 18 per cent of healthy people.[93]

The potential link between bacterial infections and digestive disorders – leading to malabsorption of B vitamins, high homocysteine, poor methylation and heart disease or strokes – is very tentative at the moment. However, it is worth bearing in mind, especially for those who have very high H scores (often above 40 units) and symptoms of both digestive problems and arterial disease, and who find little change in their H score after following the programme in this book.

The H Factor Diet and Lifestyle

20

Eat less fatty meat, more fish and vegetable protein

THERE ARE 12 GOLDEN RULES for attaining super-health, and rule number one is: eat the very best of that vital nutrient, protein. This is easy, once you realise protein is all about balance and fat content. Eating too much or too little of it spells disaster for your health.

Too much is bad because animal protein makes the body over-acid – remember, protein is made out of amino acids. This in turn taxes the kidneys and may increase your risk of getting osteoporosis. Overloading on animal protein is particularly risky if you are deficient in the homocysteine-lowering B vitamins – folate, B12 and B6. This is because animal protein is extremely rich in the amino acid methionine, which can raise homocysteine if not enough of the right B vitamins are present.[1] (See page 25 for an explanation of how this works.)

But eating too little protein is just as bad: it will mean you won't get enough amino acids. A lack of protein can even raise your homocysteine levels, according to a study of 260 retired teachers in Baltimore, Maryland. This is probably because methionine can get stuck as homocysteine if it isn't converted correctly. On the other hand, methionine is also the precursor or forerunner of SAMe, our body's primary methyl donor, which supplies methyl groups that actually help to detoxify homocysteine if the conversion goes right.[2] So your protein intake needs to stay on an even keel.

Even more importantly, you need to eat the right kind of protein. Fish, as long as it's neither fried, breaded or smoked, and lean meat are excellent sources of protein, as well as of vitamins B6 and B12, both of which keep your homocysteine at bay. Vegetables also provide significant amounts of folate and some B6, but no B12. This is why strict vegans are notorious for elevated homocysteine levels and many of the resulting complications. Hence the need for fish, lean meat, eggs and, assuming you're not intolerant or allergic, a limited amount of dairy products, supported by supplements.

The charts below show you the best food sources for B12, B6 and folate – all essential to reducing homocysteine levels.

The best foods for B12

Food	Amount per 100g serving
Sardines	28mcg
Oysters	15mcg
Cottage cheese	5mcg
Tuna	5mcg
Turkey and chicken	2mcg
Lamb	1.8mcg
Eggs	1.7mcg
Cheese	1.5mcg
Prawns	1.0mcg
Milk	0.3mcg

The optimal B12 intake from the diet is around 10mcg daily, although much more daily B12 is needed to shift a high H score into the superhealthy range.

The best foods for B6

Food	Amount per 100g serving
Salmon	0.82mg
Trout	0.68mg
Pheasant	0.66mg
Bananas	0.51mg
Turkey	0.47mg
Kidney beans	0.44mg
Mackerel	0.40mg
Haddock	0.30mg
Rabbit	0.29mg
Brussels sprouts	0.28mg

An optimal intake is 25mg per day, although again, much more B6 is needed to lower a high H score into the superhealthy range. You are unlikely to achieve more than 5mg from diet alone, which is why we recommend supplementing extra B6.

The best foods for folate

Food	Amount per 100g serving
Wheatgerm	325mcg
Lentils, cooked	179mcg
Millet flakes	170mcg
Sunflower seeds	164mcg
Endive	142mcg
Chickpeas, dried, cooked	141mcg
Spinach	140mcg
Romaine lettuce	135mcg
Broccoli	130mcg

Food	Amount per 100g serving
Kidney beans	115mcg
Peanuts	110mcg
Brussels sprouts	110mcg
Orange juice, fresh or frozen	109mcg
Asparagus	98mcg
Hazelnuts	72mcg
Avocados	66mcg

An optimal daily intake for folate is about 600 to 900mcg, but yet again, more is often needed along with other key nutrients to get a high H score into the superhealthy range. For a more comprehensive list of folate-rich foods, also look at the chart in Chapter 21.

■ Fish, the miracle food

Unlike fatty meats and dairy products, fish (and particularly oily, cold-water fish such as salmon, tuna, mackerel, herring and trout) provides the essential omega-3 fats, as well as the trace mineral selenium and the amino acid taurine, which are good for your heart, brain and arteries. Eating fish by itself apparently has no effect on raising or lowering your homocysteine levels.[3, 4] But fish eaters are much less prone to severe depression and suicide. Fish lowers blood pressure and has anti-arthritic benefits. And eating 4 to 6oz of fresh fish, cooked but not fried or breaded, a few times a week not only reduces death from heart attacks and strokes, but, as with lowering homocysteine and raising glutathione levels, reduces risk of death from all causes.

Since fish is higher is essential fats and lower in saturated fats (which do raise your H score – see Chapter 24), we give it an enthusiastic double thumbs-up. Eating more fish and less fatty meat means taking in more high-quality protein and less saturated fat.

■ Soya instead of meat

Swapping soya protein for meat protein lowers homocysteine, according to a study in which men and women with a high risk of heart disease were fed a low-fat diet, either of meat or soya protein.[5] Both diets were low in saturated fat and cholesterol. Those getting soya protein were either given soya rich in beneficial plant chemicals called isoflavones, or soya with only a fraction of the isoflavone content. Isoflavones provide some protection from breast and prostate cancer and menopausal symptoms. While compared to those eating lean meat, those eating soya lowered their H scores, and the isoflavones appeared to make no difference. We, however, recommend eating soya products that contain isoflavones (assuming you are not allergic or sensitive to soya) because of their cancer-protective effect. These include tofu, tempeh and miso.

So if soya protein is good for lowering homocysteine, how much do you need to eat to reduce your H score? A group of researchers headed by Dr Serena Tonstad from the Department of Preventive Cardiology at Ulleväl Hospital in Oslo, Norway, decided to put this to the test. They gave volunteers either 30 or 50g of soya protein a day for 16 weeks. At the end of the period both groups had had a statistically significant but modest one-unit reduction in homocysteine – without taking any other dietary measures or supplements.[6] This suggests that 30g of soya a day is a good way to keep your H score a bit lower.

A word of warning: reports of adverse reactions to soya products are on the rise in the UK, Europe and the US. For example, a 150 per cent increased incidence of allergic reactions has been reported over the last two years alone, including severe anaphylactic reactions rivalling those of peanuts. Early exposure to soya, along with wheat and milk, is thought to be an underlying cause of insulin-dependent diabetes. We suggest that you be tested for soya allergy or intolerance before making soya a major source of non-animal protein in your diet. And if you're pregnant, stick to exclusive breastfeeding in the beginning, keeping your baby off both cow's milk and soya milk for the first four months of its life, if not longer.

Although they have yet to be tested, we suspect that, like soya, other beans (such as kidney, lentil, adzuki, black-eyed and fava) will also have a homocysteine-lowering effect due to their high level of protein and low level of fat.

■ How much protein?

The guidelines are to eat a serving of lean meat a maximum of four times a week, fish a minimum of three times a week, and if you're not allergic or intolerant, have a serving of a soya-based food (tofu, tempeh, soya sausages, etc.) or beans such as kidney beans, hummus (made from chickpeas) or baked beans, at least five times a week.

What you are aiming for is two servings of protein a day, with at least one from a vegetable and one from an animal source. The chart below gives you the serving size of different sources of protein, each giving you 20g of protein. Assuming you are not a competitive or endurance athlete (they need more high-quality protein), you need, on average, two to three of these servings, that is, 40 to 60g of protein a day.

Food	Percentage of calories as protein	How much for 20g of protein	Protein quality
Grains/Pulses			
Quinoa	16	100g (3½oz)/1 cup dry weight	Excellent
Tofu	40	275g (10oz)/1 packet	Reasonable
Corn	4	500g (1lb 2oz)/3 cups cooked weight	Reasonable

Food	Percentage of calories as protein	How much for 20g of protein	Protein quality
Brown rice	5	400g (14oz)/3 cups cooked weight	Excellent
Chickpeas	22	115g (4oz)/0.66 cup cooked weight	Reasonable
Lentils	28	85g (3oz)/1 cup cooked weight	Reasonable
Fish/Meat			
Tuna, canned	61	85g (3oz)/1 small tin	Excellent
Cod	60	35g (1¼oz)/1 very small piece	Excellent
Salmon	50	100g (3½oz)/1 very small piece	Excellent
Sardines	49	100g (3½oz)//1 baked	Excellent
Chicken	63	75g (2½oz)/1 small roasted breast	Excellent
Nuts/Seeds			
Sunflower seeds	15	185g (6½oz)/1 cup	Reasonable
Pumpkin seeds	21	75g (2½oz)/ 0.5 cup	Reasonable
Cashew nuts	12	115g (4oz)/1 cup	Reasonable
Almonds	13	115g (4oz)/1 cup	Reasonable
Eggs/Dairy			
Eggs	34	115g (4oz)/2 medium	Excellent
Yoghurt, natural	22	450g (1lb)/3 small pots	Excellent
Cottage cheese	49	125g (4½oz)/1 small pot	Excellent
Vegetables			
Peas, frozen	26	250g (9oz)/2 cups	Reasonable
Other beans	20	200g (7oz)/2 cups	Reasonable
Broccoli	50	40g (1½oz)/0.5 cup	Reasonable
Spinach	49	40g (1½oz)/0.66 cup	Reasonable
Combinations			
Lentils and rice	18	125g (4½oz)/small cup dry weight	Excellent
Beans and rice	15	125g (4½oz)/small cup dry weight	Excellent

▪ Vegetarians – be careful

As far as homocysteine is concerned, being vegetarian can be bad news, unless you know what you are doing. Three studies have now shown that vegetarians are more likely to have high homocysteine levels. One survey of Taiwanese Buddhist lacto-vegetarians (who ate dairy products, but no meat or fish), compared to meat-eating Taiwanese, found the vegetarians to have the higher H scores (11.2 units for the vegetarians versus 10.5 for the meat eaters). Not surprisingly, the researchers also found lower B12 in the vegetarians.[7] As the earlier chart shows, the best food sources for B12 (one of the essential vitamins for keeping homocysteine levels low) are all animal products.

Another similar study in Spain involving 23 lacto-vegetarians and three vegans (who ate no milk or eggs) found essentially the same thing as the Taiwanese study.[8] A third study also reported low levels of the essential omega-3 fats and B12 in vegetarians, as well as raised homocysteine.[9]

The implication of these studies is not that it's wrong to be vegetarian, but rather that if you wish to be vegetarian, it is very important to ensure adequate quality protein, B12 and essential fats. As far as protein is concerned, that means eating any two or three of the 20g vegetarian protein servings listed in the chart above. In essence, that means having a significant amount of beans, lentils, nuts and seeds every day.

Supplementing with vitamin B12 (methylcobalamin is the best to take orally) is worth considering for vegetarians. Given that a diet that includes meat, fish or eggs can give you 10mcg of B12 every day, a vegetarian not eating these foods should 'top up' the recommended B12 supplement levels given on page 185 by at least 10mcg, and preferably more.

Omega-3 fats can seem to be a difficulty for vegetarians, but they're essential for healthy skin and a smoothly working immune system, brain function, co-ordination and memory. The most biologically active and health-promoting omega-3s are called EPA and DHA. They are present in fish, most wild game, and, to a small extent, in meat or eggs if the chickens are fed flaxseeds. Like the chickens, animals and fish make their own EPA and DHA from a vegetable source of omega-3 fat, alpha-linolenic acid. Cold-water plankton, the vegetable of the seas, and flaxseeds are richest in alpha-linolenic acid. For this reason, vegetarians are strongly advised to eat at least a heaped tablespoon or two of flaxseeds every day to ensure a good intake of omega-3 fats.

Most of us are not, however, very good at converting the alpha-linolenic form of omega-3 found in seeds into the biologically active forms EPA and DHA. The enzymes that do this depend on B vitamins again, plus zinc and magnesium. Hence, less strict vegetarians may be wise to supplement a fish oil capsule high in EPA and DHA to ensure they are getting an optimal supply.

Leaving moral issues aside, we believe the healthiest diet is probably a 'fishitarian' diet, basically one high in fruit, vegetables, nuts and seeds with some fresh fish (unsmoked, unfried and unbreaded), rather than meat from domesticated animals. The inclusion of some lean meat – provided it's real meat, preferably from wild game and not some intensively reared, antibiotic and hormone-fed animal – may also be equally healthy. Eating organic wild game meat and fish is the best way to ensure this. This way you are not putting any extra strain on your body's detoxification abilities by eating food contaminated with harmful chemicals, drugs or hormones.

21

Eat your greens

NOTHING TOPS FRUIT and veg when it comes to your health and maintaining a low H score. They're the richest source of folate, plus all those anti-ageing, anti-inflammatory antioxidants. All you need to do to get these fabulous benefits is eat at least five servings, or a pound in weight, of a variety of different fruits and vegetables a day. If this sounds a bit much, here's all it means in practice:

- A handful of dried apricots and figs, or a piece of fresh fruit such as an apple or pear, chopped on your (non-allergic) cereal for breakfast (one serving) (and mix in a heaping tablespoon of refrigerated flaxseed for good measure)

- An apple, pear, two kiwi fruits or an orange as a mid-morning snack, or a glass of freshly squeezed fruit or vegetable juice (one serving each)

- For each main meal make sure half of what's on your plate is fresh vegetables (two servings). A side salad is one serving

(and consider adding extra virgin olive oil as your salad dressing)

■ A cupful of raspberries, blueberries, strawberries or grapes or a large slice of pineapple or melon for desert (one serving each)

This simple – and delicious – regime can literally lower your homocysteine by at least 10 per cent in four weeks.[10]

This recommendation is endorsed by British Heart Foundation and the American Heart Association, both of whom recommend that people at high risk of heart disease meet the daily recommended intake of B vitamins (especially folate, vitamin B6 and vitamin B12).

■ Folate-rich foods

Of the three B vitamins under discussion here, folate has the most impact on your homocysteine and your health. In 1996, the US government's Food and Drug Administration (FDA) passed a law that required all enriched flour, rice, pasta, cornmeal, and other grain products to contain 140mcg of folate per 85g. This level of fortification has led to a measurable decrease in homocysteine levels. However, higher levels of food fortification with folate have been reported to be even more effective in lowering homocysteine.[11]

The recommended daily intake of folate for men and non-pregnant women is 400mcg. The recommended daily intake of folate for pregnant women is 600mcg, and 500mcg for breast-feeding women. However, assuming that other key homocysteine-lowering nutrients are included, the *optimal* amount of folate for lowering your H score into the superhealthy zone (under 6 units) is between 600 and 2400mcg per day. In practical terms, this means both eating folate-rich foods and taking supplements. Aim to achieve 400mcg of folate from your diet, topping up any additional need for other key nutrients, depending on your H score (see page 201).

Good sources of folate include wheatgerm, endive, spinach, Romaine lettuce, broccoli, peanuts, beans, seeds and orange juice, and, in the US, fortified grain products. The chart below shows you how much folate there is in a serving of different foods. Notice that a serving or two of any of these folate-rich foods is enough to get you into the optimal range of folate and is helping you maintain your homocysteine in the superhealthy zone. But remember: folate alone is not enough to lower a high homocysteine level to a superhealthy level.

Food	Amount	Micrograms (MCG)
Lentils, cooked	(½ cup)	179
Millet flakes	(1 cup)	170
Sunflower seeds	(½ cup)	164
Chickpeas, dried, cooked	(½ cup)	141
Kidney beans	(½ cup)	115
Orange juice, fresh or frozen	(1 glass)	109
Peanuts, dry-roasted	(½ cup)	106
Spinach	(½ cup)	102
Artichoke (globe)	(1 medium)	94
Miso soup	(1 cup)	91
Oats	(1 cup)	87
Asparagus	(½ cup)	80
Hazelnuts	(½ cup)	76
Romaine lettuce	(½ cup)	68
Broccoli	(½ cup)	65
Avocados	(half)	62
Brussels sprouts	(½ cup)	60
Beetroot	(½ cup)	63
Papaya	(half)	58

Food	Amount	Micrograms (MCG)
Parsnips	(½ cup)	45
Oranges	(one)	40
Cantaloupe melon	(half small)	38
Leeks	(½ cup)	32
Peas, frozen	(½ cup)	30
Kiwi fruit	(one)	29
Grapefruit juice	(1 glass)	26
Rye bread	(1 slice)	26
Sweetcorn, canned	(½ cup)	25

So here's a selection of ways to get the bare minimum of 400mcg of folate each day:

■ A salad with Romaine lettuce, endive, half an avocado and a handful of sunflower seeds, accompanied by a glass of orange juice

■ Spinach and lentil or millet bake with a serving each of broccoli and parsnips

■ A fruit salad with papaya, kiwi fruit, orange and cantaloupe melon in orange juice, plus a handful of unsalted peanuts

■ An orange, a large serving of broccoli, spinach, Brussels sprouts and a bowl of miso soup.

An even easier way of getting enough folate is simply to eat five servings of fruits and vegetables every day. Most people eat less than half this amount. In the US, less than 15 per cent of the general population eat five servings a day. In Britain, the average consumption of fruits and vegetables is around two to three servings a day, while the average intake of folate is only 239mcg.

People who eat lots of fruits and vegetables, especially dark leafy greens such as spinach, can up their folate intake from food by a further 300mcg. This extra amount is proven to lower H scores, but only by 9 per cent[12] – not enough to bring an average H score of 11 down to a safe 9, and certainly not a superhealthy 6. However, even a 9 per cent drop reduces the risk of a stroke by as much as 20 per cent. Jiang He of the Tulane University School of Public Health in New Orleans and his colleagues asked 9,764 healthy men and women to recall the foods they had eaten the previous day. Based on that survey, the researchers calculated the participants' daily consumption of folate. The team then tracked the number of strokes and heart attacks that occurred in the group over a 20-year period.

They discovered that men and women who ate a diet that provided at least 300mcg of folate per day had a 20 per cent reduced risk of stroke and a 13 per cent lower risk of cardiovascular disease, compared with people in the study who got just 100mcg of folate per day.

However, the evidence consistently shows that an intake of folate around the 100 to 200mcg a day mark just isn't enough to make a difference. You really want to achieve well above 400mcg to help the other homocysteine-busting nutrients keep homocysteine at bay. We recommend 900mcg as your base. That's why we say to eat at least five servings of fruits and vegetables every single day, but don't forget to supplement with extra folate!

▪ Don't overcook your vegetables

Your mother might have told you to boil peas for five minutes and runner beans for seven. Sad to say, this is all wrong. Overcooking destroys 90 per cent of the folate in food and inevitably raises your homocysteine level. Most vegetables should be cooked, if at all, for the shortest amount of time possible, until hot but still crunchy. It should be noted that the nutrients in some vegetables, such as carrots, leeks and tomatoes, do become more available by light cooking. Steaming or stir-frying are better methods than boiling, as the nutrients don't leach away into the water.

Here's a guideline of how long it takes to get your vegetables hot and still crunchy, with some folate intact:

Vegetable	Cooking time
Broccoli/cauliflower	3 minutes
French beans	3 minutes
Spinach	1 minute
Cabbage/greens	2 minutes

Many cultures have the habit of overcooking vegetables – a leftover from necessary food hygiene practices to kill off nasty microbes. The trouble is that such overcooking can kill you off, too, by destroying all the folate in food and raising your homocysteine to a dangerous level. Asian semi-vegetarian diets, for example, are often very low in folate and other vitamins and minerals lost from overcooking,[13, 14] and many millions of Asians suffer from high homocysteine as a result.

■ Eat right *and* take B vitamins

There's a hitch to all this, though. Research also clearly shows that folate from food alone is not nearly as effective as folate from supplements (see page 194). Dr Manuel Malinow, who has done extensive research on the effects of feeding people different diets fortified with folate, believes, 'We as doctors need to change from saying "Eat right or take a multivitamin" to saying "Eat right and take a multivitamin." Folate from fortified grain products is more readily absorbed by the body than naturally occurring folate in foods, so fortification is very important.' We would only add that folate supplements alone are very important but are not enough. You need to take all the key homocysteine-lowering nutrients.

22

Have a clove of garlic a day

GARLIC MAY QUELL romance, but in true matters of the heart – the health of this all-important muscle – it's invaluable. Many ancient civilisations knew this. Way back in the first century, the Romans believed that garlic 'clears the arteries'. And even further back, the Indian Ayurvedic text *Charaka Samhita* states that garlic 'maintains the fluidity of the blood, strengthens the heart and prolongs life'.

Now, modern research has backed up these ancient beliefs by showing that raw and cooked garlic and onions contain compounds that inhibit the tendency to form artery-clogging blood clots, thus lowering heart attack and stroke risks. There is also some evidence that a clove of garlic a day may lower high levels of blood cholesterol, as well as lower high blood pressure.

Other 'allium' compounds – the active component in garlic, onions, leeks, chives and shallots – have been shown to inhibit cancer in the test tube and in animals. In population studies, people who eat more of these vegetables have lower rates of stomach and other cancers. In a major five-year study in Iowa of more than 35,000 women aged 55 to 69, garlic was the single most protective food against colon cancer. The women who ate garlic more than once a week had a 32 per cent lower risk of colon cancer than those who ate less than a clove of garlic a month. Garlic's sulphur compounds, from which the body can make the most powerful antioxidant of all, glutathione, may offer particular protection against cancers of the breast, oesophagus, prostate, skin and stomach.

But why is garlic so good? First of all, garlic and other allium vegetables are naturally rich in the sulphur-containing amino acids that help make glutathione. Garlic also appears to stimulate the body's production of an enzyme called glutathione-S-transferase, which helps detoxify potential carcinogens. It's also associated with general detoxification, and as we have learned, when glutathione levels increase, homocysteine levels are lowered.

Whether or not this alone is the reason, there's no doubt that garlic is an excellent food for lowering your H score, according to the results of a trial with some very smelly rats!

Dr Yu-Yan Yeh and his team of researchers at the Pennsylvania State University put young rats on a diet low in folate. Not surprisingly, they developed high homocysteine. They were then given aged garlic (which is more potent). Those that had the highest H scores had up to a 30 per cent reduction! Those with lower H scores still had a reduction, proving that garlic does lower your homocysteine – which may be yet another reason why it's good for the heart.[15]

That's why we say 'a clove of garlic a day keeps the doctor away'. It's essential. This means that when you are making dinner, you need to add a clove of garlic per person to the main dish. This is what southern Mediterraneans do. But cooking decreases the potency of the garlic, so it's best to crush it and stir it in just before serving.

If you really don't like the taste or smell, you can take a garlic capsule instead. Some are de-odorised.

23

Don't add salt to your food

SALT BRINGS OUT the flavour in food ... but the bitter truth is that it can be highly dangerous. One in ten people are 'salt sensitive', for instance, meaning that their blood pressure goes up when they eat food high in salt. This increases their risk of heart disease.

However, all of us may be vulnerable to a hidden danger in salted foods. Recent research in animals has shown that if you have both a high homocysteine level and a high salt intake, the inner lining of the arteries become much more damaged, indicating the first stages of heart disease.[16] Interestingly, this damage was not seen in those who had low homocysteine levels while consuming a lot of salt.

So, unless you have an H score below 6, we strongly recommend that you cut back on salt, both by not adding it to food, and by choosing food without added salt. Or, since sodium chloride, or table salt, is the culprit here, you could go for a special salt with beneficial properties. This is Solo Salt, a natural sea salt with 46 per cent less sodium and a lot more of the minerals magnesium and potassium instead. This type of salt has been shown to lower, not raise blood pressure.[17]

■ How much sodium is there in your food?

Foods are listed in order of those that contain the most sodium per calorie of food. The figures in brackets show the amount of sodium in 100g, which is roughly the equivalent of a cup or serving.

Shrimps	(2950mg)	Crab meat	(369mg)
Green olives	(2300mg)	Tuna fish	(339mg)
Parmesan cheese	(1862mg)	Kidney beans (canned)	(327mg)
Bacon	(1603mg)	Cream cheese	(300mg)
Ham	(1500mg)	Mussels	(286mg)
Sardines	(650mg)	Beef liver	(184mg)
Cheddar cheese	(622mg)	Egg	(138mg)
Low-fat milk	(549mg)	Salmon	(116mg)
Cottage cheese	(405mg)	Cod	(109mg)
Whole milk	(371mg)		

24

Lower your fat

IT'S BEEN DRUMMED into us for years, but certain kinds of fat – saturated or damaged fats, mainly from meat, dairy products and fried or processed foods – aren't good for us in excess. They raise your homocysteine level, affect heart health, can lead to obesity . . . in short, they are a major dietary bugbear.

But you can't throw all fats into the same basket. Some are absolutely essential to health. Appropriately, they're called essential fats, and as you can't manufacture them yourself, you need them in your diet on a regular basis. We'll be discussing them below.

And you will need to keep the amount you eat within healthy limits. In the West, the average intake of fat is 40 per cent of our total calories. The amount should ideally be less than 30 per cent, composed of good, essential fats.

So what's the ideal way to ensure you're getting the right kinds and proportions of fats? It means eating:

■ Less (domesticated) meat, and the meat you eat should be lean

- Less cheese and fewer dairy products

- Fewer fried foods

- Fewer high-fat, processed foods loaded with hydrogenated oils

- More fresh oily fish (salmon, tuna, mackerel, herring, trout, orange roughy)

- More fresh, unsalted nuts and seeds.

Dr Fleming from the University of Nebraska Medical Center in the US decided to see what difference following such a low-fat diet would make to homocysteine, compared to those eating high-fat diets. After a year, he found that the low-fat diet lowered H scores by 14 per cent, while the high-fat diet raised H scores by 10 per cent.[18]

This is why we recommend meat no more than four times a week, and very lean meat at that. It is also best to stay away from cheeses high in saturated fat, such as those shown below.

How much saturated fat is there in cheese and meat?

Food	% of calories as saturated fat
Gruyère	71
Processed cheese	73
Cheddar cheese	74
Roquefort	75
Sirloin steak	70
Leg of pork	72
Ground beef (regular)	77
Pepperoni	81
Leg of lamb	82
Pork sausage	87
Bacon	93

Remember, it's best to eat less than 30 per cent of your total calories as fat, which means eating foods that are less than 30 per cent fat. Some lean meat products are, but you have to read the label carefully, even if a product claims to be low fat.

So how do you know for sure what percentage of fat is in the food you buy? Here is a simple equation to help you.

Test your food with the fat test

Look at the number of calories per 100g on the label. Now look at the number of grams of fat per 100g and multiply it by 10. Is this more than a third of the number of calories per 100g? If so, it's more than 30 per cent fat.

For example, yoghurt may provide 60 kcals per 100g. The fat content per 100g is 3.5g. Multiply by 10, giving you 35. Divide 35 by 60 (0.58). This means that more than 50 per cent of the calories in this yoghurt comes from fat.

But remember that you need to concentrate on essential fats. Here, balance is vital.

Of the two families of essential fats – omega-6 and omega-3 – most of us are severely deficient in the omega-3s. These, as we discussed in Chapter 20, are only found in fish, and particularly oily cold-water varieties, nuts and flaxseeds, and to a lesser extent in wild game and other seeds. We need equal proportions of omega-3 and 6, which is exactly what our pre-agriculture ancestors used to eat for hundreds of thousands of years. Today's diets have 20 times more omega-6 than 3, and are deficient in both. The main reason for these very low levels of omega-3s in our diet is that we've stopped eating fish and wild game. Another reason is that these fats go rancid quickly. Hence, all processed foods with long shelf-lives are devoid of omega-3s. Good foods tend to go off: it's a fact of life.

So while recommending you to lower your saturated fat from meat and dairy, hydrogenated fat from processed foods, and damaged fat from frying and overcooking, we recommend you eat more oily and other fish, nuts and seeds, and use cold-pressed seed oils in salad dressings as a source of essential fats.

25

Cut back on tea and coffee

THE CUP THAT CHEERS? Only in moderation. So if you're still reaching for the coffee first thing in the morning, punctuating your workday with cups of tea, and meeting friends over a tall latte or two, it's time to think again. Coffee and tea are bad news as far as homocysteine is concerned, and you will need to limit them.

For instance, according to research from the Netherlands, just two weeks of drinking coffee can raise your H score by 10 per cent. A group of doctors from the University Hospital Nijmegen tested the effects of coffee by assigning volunteers to drink a litre of unfiltered coffee a day – that's about four cups – for two weeks. At the start of the two weeks their average H score was 12.8 units, slightly above the national average of 10 to 11. At the end of the two weeks their H score was 14 units.[19]

A more recent study by Dr Petar Verhoef and colleagues at the Wageningen Centre for Food Sciences in the Netherlands showed that two cups of regular coffee increased homocysteine by 11 per cent after only four hours, while caffeine tablets without coffee increased it by only 5 per cent.[20] Clearly it's not just the caffeine in coffee but other ingredients as well. So don't rely on decaffeinated coffee. It may be bad for your homocysteine level, too.

■ The truth about tea

At this point you may just think you'll switch to tea. We've all been reading the good health reports about black and green teas, haven't we? But be aware that the main culprit in coffee is thought to be something called chlorogenic acid, a type of chemical compound called a polyphenol. Most 'phenols' are toxic to the body; you might have heard of PCBs or polychlorinated-biphenol, for example, one of the most widespread environmental pollutants. Regular black tea is also high in polyphenols.

To test if it was chlorogenic acid or the other polyphenols in coffee that raise levels of homocysteine, another group of Dutch researchers – this time from the Division of Human Nutrition and Epidemiology at Wageningen University – devised an experiment. They gave some volunteers 2g of chlorogenic acid, which is about what you get in 2 litres of coffee. They gave other volunteers 4g of black tea, which gives about the same amount of polyphenols that you'd get in 2 litres of tea, and others dummy pills. Each group did this for a week, and then was randomly switched over for the next week, and then for a further week. In other words, a volunteer could be on chlorogenic acid for one week, then placebo, then black tea. In this way they were able to really find out what the effect was.

Homocysteine went up 12 per cent after chlorogenic acid, and 11 per cent after black tea.[21] In other words, both tea and coffee raised homocysteine levels.

The link between high coffee consumption and high H scores has been found in other studies investigating which dietary factors seem to make the most difference.[22]

So we recommend that you drink no more than one or two regular cups of strong tea or coffee a day. After you've reached the limit, go for herbal teas or coffee substitutes. (Remember, decaffeinated coffee still contains chlorogenic acid and therefore has the potential to raise cholesterol, so we recommend you avoid this as well.)

You might think that this condemns you to drinking wishy-washy 1970s-style brews. On the contrary: delicious alternatives to tea and coffee abound these days. Herbal and fruit teas, for instance, come in a huge, inventive range:

- Red Bush (Rooibos) – tea that tastes a lot like regular tea but contains no harmful chemicals and is rich in antioxidants

- Fruit teas – such as blackberry, lemon, raspberry, cranberry or strawberry

- Herb teas – such as mint, camomile or fennel

- Coffee alternatives – such as Caro, Bambu and dandelion coffee.

All should be available from good supermarkets and health-food shops.

26

Limit your alcohol

DON'T PUT THAT glass down yet. Depending on your favourite tipple, you may be able to have an evening drink without upsetting your H score. What's certain is that too much alcohol is definitely bad news for your homocysteine levels.

Concern over alcohol's effect on homocysteine levels originally arose after surveys such as the American Framingham study found that the more alcohol consumed, the higher the H score.[23] However, how much is too much is an interesting question. To test this, another group of researchers gave 60 normal, well-nourished people 30g of alcohol a day, the equivalent of 1.5 pints of beer, three regular glasses of wine, or three measures of spirits. At the end of the six weeks, whatever form of alcohol they'd been drinking, their homocysteine score was higher.[24]

But not all research gives all forms of alcohol the thumbs-down. Some recent studies conclude that beer in moderation, probably due to its higher B vitamin content, can actually lower homocysteine! Others conclude that red wine is better for you than beer. So, what is the best form of alcohol as far as your H score is concerned?

■ Beer versus wine

According to Dr Thomas Truelsen, from the Institute of Preventive Medicine in Copenhagen, Denmark, people who drink wine every week are 70 per cent less likely to develop dementia after the age of 65 than those who don't drink wine. But those who drink beer, even as infrequently as once a month, are twice as likely as those who don't drink beer to develop dementia after the age of 65.[25] The researchers came to this conclusion after collecting detailed information from almost 2,000 people 15 years ago about what they drank and how often they drank it. Then, when all the participants were at least 65 years old, the researchers returned to find out if they had developed dementia and then compared this information to past drinking behaviour.

But what do beer and red wine do to your H score? A team led by researchers at the TNO Nutrition and Food Research Institute in the Netherlands decided to find out. They gave lucky volunteers red wine or beer to drink, an average four glasses a night (the equivalent of 40g of alcohol) for 12 weeks. At the end of this time, while the scores had gone up for red wine, they'd actually gone down for beer![26]

And this team is not the only one to find that beer lowers homocysteine. A recent study from Pilsen in the Czech Republic gave over 500 volunteers a free litre of beer a day. (They had no problem getting recruits!) Once again, they found that this amount of beer was associated with slightly lower H scores, and higher folate and B12 levels in the volunteers' bloodstreams.[27] Why? They speculated that the B vitamins in beer might have had a protective effect.

But before you rush out to your local pub for a pint, another study, this time from England, compared the effects of beer, wine and spirits in different amounts on the homocysteine levels of 350 obese people. They found that those consuming the equivalent of a drink a day had slightly lower H scores than those who abstained (8.5 versus 9.5 units). They also had slightly higher folate levels, again suggesting that their moderate drinking was a source of B vitamins. However, in this study, beer came out badly, while red wine came out on top. The red wine consumers had an average score of 7.8, the white wine consumers 8.8, the beer and spirit consumers scored 9, while the abstainers scored 9.4.[28] Their conclusion is that drinking around 100g of alcohol from red wine a week, or a glass of wine a day, may slightly lower your H score.

Confused? You should be. But science is sometimes like this! No doubt the beer and wine war will continue with further research. One thing there is no confusion about is the harmful effects of drinking more alcohol than this, especially for women. In fact, the evidence so far in relation to breast cancer suggests that even a glass of wine a day may be bad news.

■ Women, alcohol and breast cancer

A recent report in the journal *Cancer* found that among women who had a mother or sister with breast cancer, those who drank daily had almost 2.5 times the risk of breast cancer than those who did not drink alcohol at all. Another study reported that women whose diets were lowest in folate faced no greater cancer risk than women with higher-folate diets – if they were non-drinkers. But if they drank more than two alcoholic drinks a week, their breast cancer risk increased almost 60 per cent! The *Journal of the American Medical Association* earlier reported similar findings from a large-scale survey of female nurses. They found that the risk of breast cancer increased 40 per cent among women who consumed the least folate and drank one or more drinks a day. Those who drank the same amount, but consumed folate near or above the recommended 400mcg a day faced no significant increase in risk.

How can we make sense of all these studies? Well, one clear suggestion is that getting enough of the homocysteine-lowering B vitamins is good news. Possibly, for those who are deficient, the amount provided in a beer or a glass of wine could be slightly beneficial and offset alcohol's negative effects. Also, having an optimal intake from diet and supplements may also render alcohol less toxic. However, it's a fine line, and moderation is the key word.

We therefore recommend you drink no more than four glasses of red wine or beer a week, *provided* your homocysteine score is below 9. However, if you have a high H score, we recommend you avoid it completely until you are, at least, down to this level.

■ Alcohol and homocysteine damage your brain

Having a high homocysteine score can directly damage the brain, both by knocking out brain cells and by increasing levels of free radicals, harmful chemicals that also attack the brain.[29] It is very likely that high homocysteine, which can be induced by excessive drinking, accounts for the brain-damaging effects of alcohol.

But what about the liver, the other organ that suffers badly from excessive alcohol? Again, the destructive agent may be homocysteine. Remember that enzyme, MTHFR, that helps make SAMe, the body's most important methyl donor, from homocysteine (see page 25)? This is also the enzyme that fails to work very well in about one in ten people. Alcohol impedes its action further still, effectively starving the body and the liver of the products of optimal homocysteine metabolism (methyl groups, SAMe[30] and glutathione). The net result is that the liver becomes more and more susceptible to damage and life-threatening diseases like alcoholic cirrhosis.

This discovery is very important for the treatment of alcoholism and the damage caused by alcohol because it shows that supplementing SAMe, or even better TMG (a more readily utilised supplement containing three methyl groups), can not only prevent but undo the damage caused by alcohol.[31, 32]

In a nutshell, what all this means is that the worst thing to do is drink a lot of alcohol while being deficient in all the key H Factor nutrients (B6, B12, folate, B2, zinc and TMG), especially if you are one of the 10 per cent of people who have a faulty MTHFR gene.

Conversely, if you do have a drinking problem, the best thing you can do is to take all these nutrients up to effective levels, by following the programme in this book. Although we shouldn't be saying this, this homocysteine-lowering strategy, with maximum amounts of TMG, is also the best cure for a hangover. You'll see more on this in Part 5.

27

Keep cool and reduce your stress

STRESS SEEMS TO BE a mainstay of 21st-century life. A certain level is good, as it keeps us alert and able to cope with the obstacle course of daily living. But if you're overloaded with stress, it's very bad news. Excessive stress has long been known to increase risk of a heart attack. In fact, the archetypal heart attack as portrayed in Hollywood movies is usually stress-induced. Moreover, stress is now known to be associated with inflammation in the walls of arteries, and elsewhere in the body, which is the body's 'red alert' signal.

And, although the exact mechanism isn't known, stress raises your level of homocysteine.[33]

Stress associated with hostility and repressed anger[34] is particularly dangerous and raises homocysteine levels. If you find that you're excessively grumpy, critical, bad-tempered or thinking angry thoughts, don't be surprised if your H score is high. Check yourself out on the stress test below.

Test your stress

☐ Is your energy less now than it used to be?

☐ Do you feel guilty when relaxing?

☐ Do you feel a persistent need for achievement?

☐ Are you unclear about your goals in life?

☐ Are you especially competitive?

☐ Do you hate to lose?

☐ Do you hate being corrected, lectured or scolded?

☐ Do you work harder than most people?

☐ Do you easily become angry?

☐ Have you been in a physical fight since high school?

☐ Are you irritable much of the time?

☐ Do you often do two or three tasks simultaneously?

☐ Do you get impatient if people or things hold you up?

☐ Do you have difficulty getting to sleep?

A score of five or more puts you in the high stress category.

Of course, it's one thing to say 'keep cool' and another to do it. Having said that, there are highly practical ways you can chill out. For starters, following the H Factor programme should reduce your level of stress. This is because a superhealthy diet will leave you feeling energised yet calm, while munching on sugar, hydrogenated fats or allergic foods, drinking stimulants or alcohol, and failing to take any B vitamins is a classic formula for upping your stress level.

In a survey conducted by the Institute for Optimum Nutrition in London, which involved giving people 'optimum nutrition achieved with dietary changes plus supplements', 79 per cent noticed an improvement in energy, 60 per cent had better memory and mental alertness and 66 per cent felt more emotionally balanced. At the beginning of the survey over half of the people, 54 per cent, rated high on a simple stress check (shown above), scoring five or more 'yes' answers. After six months on this optimum nutrition programme, with no other stress-reducing therapy involved, the number with a high stress rating had dropped to only 28 per cent. Simple nutritional changes along with supplements almost halved the number of high stress scorers. Yet on the surface, these questions appear to have nothing to do with nutrition.

There are many other ways you can reduce your stress levels. Here's a few examples:

- Practise deep breathing. When you're stressed out, one of the quickest ways to regain balance is to breathe deeply and gain some perspective on whatever is making you feel anxious. Breathing deeply also brings more oxygen into the body, and consequently more energy.

- Ensure you consume adequate amounts of magnesium. Chronic stress depletes your body of magnesium and the more stressed you are, the more magnesium you lose. The lower your magnesium level to begin with, the more reactive to stress you become, and the greater the loss of magnesium from cells. Increasing your intake of magnesium through diet and supplements (a chelated magnesium like magnesium glycinate is preferred) breaks this vicious cycle by raising blood magnesium levels and helping you deal better with stress. So eat pumpkin seeds, green leafy vegetables and whole grains such as brown rice and, assuming you're not gluten sensitive, rye and wholewheat breads.

- Take regular exercise. Exercise is an essential part of managing stress. Our evolutionary stress response – known as the 'fight-or-flight mechanism'– is designed to give you a sudden burst of energy to save your life if, for example, a lion is chasing you. Most of us are now safe from lions, but we have 21st-century stresses – work, money, relationships, world news, city living – all day, every day, and our bodies still react in the same way, preparing us for intense physical activity. Exercise uses up that energy, stopping it from building up as tension and anxiety in your mind and body.

- Meditate every day for ten minutes. Meditation is the ultimate antidote to the effects of stress on every level. Physically, it reduces heart rate and blood pressure, slows the rate of breathing and stabilises brain wave patterns. It also improves the body's responsiveness to stressful events and aids quicker recovery and has also been shown to prevent the depression of the body's immune responses that occur with stress. (See page 262 for details of meditation courses.)

28

Stop smoking

AS WE ALL KNOW by now, smoking is very tough on the body. And as far as your H score is concerned, it's one of the worst things you can do, making it a major drain on your overall health. Homocysteine is not only higher in those who smoke, it's also higher in children whose mothers smoke. If you're a heavy smoker, the chances are very high that you have an elevated H score, along with the non-smokers living with you: research shows that H scores are, on average, 18 per cent higher in those who smoke 20 or more cigarettes a day.[35]

Smoking is notoriously difficult to quit and many smokers manage to reduce to less than 10 a day, hoping that will reduce risk of disease down the track. Dr James Stein and colleagues from the Preventive Cardiology Program and Center for Tobacco Research and Intervention at the University of Wisconsin Medical School in the US wondered to what extent cutting back would reduce homocysteine. So they devised a study to find out. They took 51 'healthy' smokers who smoked, on average, 35 cigarettes a day. They were then assigned to one of three groups:

those who continued to smoke; those who reduced their smoking to four to eight cigarettes a day; and those who quit – and were given help to do it. As you might expect, those who quit had a 12 per cent reduction in their H score. However, those who reduced their smoking to just four to eight cigarettes a day but did not stop showed no significant change.[36]

So it's vital to stop smoking, and that includes cigars and pipes. Cutting back just isn't enough to make a difference to your homocysteine, although it's obviously a step in the right direction as far as health is concerned.

It is doubly important not to smoke during pregnancy because smoking knocks out folate, B12 and other vitamins, as well as raising homocysteine.[37] Since smoking, high homocysteine, low folate and low B12 are all associated with increased pregnancy problems and risk of birth defects, it is strongly recommended that any woman wishing to get pregnant first stops smoking (and don't forget to get your homocysteine level tested).

29

Don't diet without supplements

IT'S OBVIOUSLY A GOOD thing to be fit and the right weight for our build. Yet two studies have shown that losing weight raises homocysteine! Before you think you might as well give up your diet, let us explain.

The first study measured homocysteine levels in 293 obese people who had 'lap band' surgery, which involves reducing the stomach size by putting a ring around the digestive tract. This relatively minor surgical procedure means people feel fuller faster and for longer, and so eat less.

One year after the surgery, the researchers found that those who lost the most weight had the highest homocysteine levels, despite little change in blood levels of folate and B12.[38] (Keep in mind here that they were probably deficient in folate and/or B12. Homocysteine levels are always better indicators of folate and B12 nutritional status than blood levels.) Those who lost weight and took B vitamin supplements didn't have the same increase in homocysteine. This was the first hint of evidence that weight loss itself might raise homocysteine, and that homocysteine-lowering supplements might be an essential addition to any weight loss diet or programme.

The chances are you're not about to have 'lap band' surgery, so is this of any relevance to you? The answer is 'yes', because the same high homocysteine level has also been found as a consequence of losing weight on a conventional calorie-restricted diet. In this trial from University Klinik Marienhospital in Germany, overweight people were put on a calorie-controlled diet. Some also took supplements of B6, B12 and folate, while others did not. Once again, those who lost weight but didn't take supplements had almost a one-unit increase in their H score, while those on the supplements had a decrease of almost one unit.[39]

These two studies both imply that losing weight is bad news for your homocysteine level and puts extra stress on the body. To what extent this is because diets tend to be lower in essential homocysteine-busting nutrients or because losing weight increases the need for these nutrients is not yet known.

However, the advice is clear. Don't go on a diet without knowing your H score, and following the H Factor supplement recommendations in Part 5.

30

Correct oestrogen deficiency

WOMEN ARE LUCKY in many respects: the female hormone oestrogen, for instance, has a highly protective effect on health. Women die of heart attacks far later than men, and one of the reasons for this is thought to be their higher levels of oestrogen up to the menopause. After this watershed, their risk of heart disease starts to rise rapidly to the same level as men. But how oestrogen protects against heart disease has, until recently, been a mystery – as has homocysteine's part in the drama.

Two recent studies examining the connection between H levels and heart disease show that oestrogen lowers blood levels of homocysteine. One, a study from Peking's University Hospital, found that six weeks on low-dose oestrogen hormone replacement therapy (HRT) resulted in a 14 per cent decrease in homocysteine.[40] Another found an 11 per cent decrease when low-dose oestrogen HRT was given to elderly men.[41]

The reason why oestrogen may help keep homocysteine at bay is that it enhances the activity of an enzyme, TMG, that helps turn harmful homocysteine into its extremely helpful cousin, SAMe. However, before you rush out and start taking oestrogen HRT, there are some things you need to know. First, while being on HRT may decrease homocysteine, there is little evidence that it actually does much to reduce risk of heart disease. Four studies involving 20,000 women on HRT for an average of five years found no difference in heart attack incidence compared to those not on HRT, but did find an increased incidence in strokes.[42]

What's more, the now well-established increased risk of breast and uterine cancer conferred by oestrogen and progestin HRT makes it impossible for us to recommend it to lower your homocysteine levels. In case you're not convinced of the danger, see the results of these four studies below:

- HRT for longer than five years doubles the risk of breast cancer, and the risk is higher if oestrogen plus progestins are given.
  L. Bergvist et al., *New England Journal of Medicine*, vol. 32 (1989), pp. 293–7.

- HRT for five or more years increases breast cancer risk by 71 per cent, and the risk is higher if oestrogen plus progestins is given.
  G. Colditz et al., *New England Journal of Medicine*, vol. 332 (1995), pp. 1589–93.

- Ovarian cancer risk is 72 per cent higher on oestrogen HRT.
  C. Rodriguez et al., *American Journal of Epidemiology*, vol. 141, no. 9 (1995), pp. 828–35.

- An Oxford University review of all research up to 1997 concluded that 'HRT raises the risk of breast cancer by 25%'.

> ■ Combined oestrogen and progestin HRT for five years
> increases risk of invasive breast cancer by 26 per cent,
> strokes by 41 per cent and heart disease by 22 per
> cent.
> Women's Health Initiative, *Journal of American Medical Association*, vol. 288,
> no. 3 (2002), pp. 321–33.

So oestrogen HRT may lower homocysteine, but at a considerable cost. The question is, how can you ensure optimal oestrogen without the risk?

Thanks to the excellent work of Dr John Lee, an expert in natural hormone replacement, we now know that the risk of breast cancer associated with HRT is due to 'oestrogen dominance' – that is, too much oestrogen, unopposed by its balancing partner progesterone. We also know that progestins, which are messed-up molecules that vaguely resemble the body's own natural progesterone, increase this risk, while natural progesterone reverses it. All this is well explained in Dr Lee's excellent book, *What Your Doctor May Not Tell You About Breast Cancer*, co-authored with David Zava and Virginia Hopkins (Thorsons, 2002).

We also believe that many oestrogen-deficient postmenopausal women have B vitamin deficiencies, zinc and magnesium deficiencies, methylation deficiency, SAMe, nitric oxide and glutathione deficiencies, and high homocysteine levels.

So how *do* you ensure optimal oestrogen, balanced with progesterone? Our advice is if you are post-menopausal, you should see an informed doctor, or, failing that, a nutritional therapist. They can test your hormone balance using a saliva hormone test – which is much more accurate than a blood test – and then recommend natural hormones, such as natural progesterone, possibly in conjunction with low-dose oestrogen. They can then retest you down the track to make sure everything is in balance.

By the way, the body makes its own oestrogen directly from progesterone. So supplementing natural progesterone, given as a transdermal skin cream, can often correct oestrogen deficiency with no possibility of oestrogen overload.

For details on how to find an informed doctor, contact the Natural Progesterone Information Society (details on page 263). To find a nutritionist, see page 263.

To sum up: we advise post-menopausal women in particular to get their hormone levels checked and, if low in oestrogen and/or progesterone, correct this with natural HRT using natural progesterone (not progestins) and, if necessary, small amounts of oestrogen. When you get your hormones checked and before you begin HRT therapy, have your homocysteine level checked as well. If your H Factor is high and your hormones low, the best therapy would be to treat both with natural HRT and homocysteine-lowering supplements, which you'll find out about in Part 5.

31

Avoid certain medical drugs

A RELATIVELY SMALL number of prescribed medical drugs have been tested and found to raise homocysteine. These include:

- Methotrexate[43]

- Fibric acid derivates used in diabetes[44]

- Theophylline used for asthma[45]

- Corticosteroids, used in inflammatory diseases including asthma and arthritis[46]

- Sulphasalazine, used for arthritis[47]

- Phenytoin, phenobarbital (PB) and primidone used as anti-convulsants in epilepsy[48]

- Cyclosporin

- L-Dopa, used for Parkinson's disease

- Metformin, used in diabetes[49]

- Cholestyramine, used to lower cholesterol

- Protease inhibitors used for HIV and AIDS.

Some of these drugs are bad news as far as homocysteine is concerned because they knock out folate. If you need to be on them, make sure your doctor is also giving you sufficient folate supplement. If you follow our recommended H Factor supplement programme in Part 5, this will help minimise the effects of these drugs on your H score.

PART 5

The H Factor Supplements

32

Eat right *and* take supplements

WITH A DIET and lifestyle geared to keeping your H score low, you're all set for superhealth. Almost: there is one more ingredient in this magic mixture, and that's daily nutritional supplements.

As we've said, just following the diet-and-lifestyle component of the H Factor programme is good, but unlikely to lower already high homocysteine to a superhealthy level. Nor would food and lifestyle changes alone be nearly enough if you have the genetic variation of the MTHFR enzyme (which makes one in ten people in the USA, UK and Europe prone to high homocysteine levels). It's only by adding in daily supplements of certain key nutrients and herbs that you can both dramatically lower your H score in half the time, and promote your health literally from day one.

As we've seen throughout the book, the most important nutrients that help lower your homocysteine and improve your 'methyl IQ'– your body's ability to keep everything is balance – are:

- Folate

- B12

- B6

- B2

- Zinc

- Trimethylglycine or TMG.

On the basis of all the research published to date, we've worked out approximately how much of each nutrient you need every day, depending on your H score. This is shown in the table below. Notice that as your H score drops closer to and is maintained at the superhealthy range, your daily need for these supplemental nutrients also decreases accordingly.

Nutrient (dose/day)	No risk Below 6 units	Low risk 6–9	High risk 9–15	Very high risk Above 15
Folate	200mcg	400mcg	1,200mcg	2,000mcg
B12	10mcg	500mcg	1,000mcg	1,500mcg
B6	25mg	50mg	75mg	100mg
B2	10mg	15mg	20mg	50mg
Zinc	5mg	10mg	15mg	20mg
TMG	500mg	750mg	1.5–3g	3–6g

Just in case you are thinking this all sounds very complicated, a number of supplement companies already produce combination supplements that give you all these in one or two capsules or tablets (see 'Directory of Supplement Companies' on page 264). If you have a very high H score, this means you might start by taking three tablets, twice a day. But, once your level hits the superhealth category and you are able to maintain it there, confirmed by retesting your H score, two a day will probably suffice to keep you safe.

In the right combination and doses, multiple nutrients are far more powerful and consistent at lowering your H score than any one in isolation. This is the principle of synergy, which we'll discuss in Chapter 37. It works because each of these nutrients is a vital piece in the jigsaw of how your body keeps homocysteine at bay.

B vitamins are water soluble, and leave the body in a matter of hours. So, if you are taking two or more of the same homocysteine-lowering supplements a day, you will derive more benefit by taking, say, one in the morning and one at lunch or early dinner.

B vitamins are usually absorbed well with or without food. One critical exception is vitamin B12, commonly given as hydroxycobalamin or cyanocobalamin. Many adults, especially as they age, have trouble absorbing these forms of B12 when taken orally, and may need vitamin B12 injections periodically. In such cases, as an alternative to injections we recommend the methylated, activated oral form of vitamin B12, methylcobalamin, which is extremely well absorbed and highly effective.

TMG is best absorbed on an empty stomach. The same is true for zinc and magnesium. So the ideal is to take these on rising, at least 20 to 30 minutes before breakfast.

■ Your homocysteine level predicts your ideal nutrient intake

So far, we've looked at B vitamins and other key nutrients as a way of lowering your H score. But you can also look at the relationship the other way round – that is, your H score can predict your optimum requirement for B vitamins. In the same way that everyone's appearance and genetic make-up differ, everyone has a different requirement for individual nutrients, and none more so than B vitamins. The important question then becomes, how do we monitor and measure our individual need for B vitamins?

When homocysteine first arrived on the medical map in the

1970s, it took some time to realise that the link between homocysteine and B vitamin status was very strong. More recently, homocysteine has come to be seen as a useful, highly sensitive indicator of deficiency of folate and B12 status, and possibly B6 – even better than the conventional blood tests used by doctors and nutritionists. As more and more studies confirmed that low B vitamin status meant high H scores,[1, 2, 3] we started to realise that your H score is the best single indicator of your B vitamin status.

The reason for this is that your H score actually tells you whether the key B vitamins in your diet are doing their job, converting homocysteine into the good guys, SAMe or glutathione. So it is a dynamic 'functional' test, and the best way to discover whether your body's homocysteine chemistry is actually working as well as it can.

Before homocysteine came along, a person's need for additional B vitamins was often tested simply by looking at dietary intake levels or blood levels. But dietary levels are unreliable, since some people don't recall accurately what they eat, the B vitamin content of the food eaten is not known with precision and varies even in the same foods, and many people's ability to absorb nutrients varies. People with chronic digestive problems, such as coeliac disease, for example, may seem to be getting enough B vitamins, magnesium or zinc from their diet, but still be chronically deficient. And while blood levels of B vitamins, magnesium and zinc may be useful at times, they still don't actually tell you if these key nutrients are activating homocysteine-lowering enzymes inside your cells in an optimal way or not.

Take this example: Fred has a high H score, but a 'normal' B6 blood level. Yet, when he supplements daily with 100mg of B6, his H score goes down substantially and his B6 blood level goes up. What does this tell you? Well, when Fred has more B6, he becomes healthier by lowering his high H score. So he was insufficiently nourished (lack of efficiency in lowering homocysteine) with B6, even though his blood level was 'normal' to start with.

The conventional B6 blood test didn't pick this up; the homo-cysteine test was a much better indicator.

That's why we recommend you supplement the big four B vitamins if your H score is high, even if your dietary intake or blood levels of B vitamins are 'normal'.

■ What is normal?

Also, bear in mind that 'normal' is a loaded, often misleading word. A 'normal' H score in the UK and the US, remember, is 10 or 11 units. Many uninformed doctors would tell you exactly this, implying that this is where your homocysteine should be. What an H score of 10 or 11 units actually means is that you are in suboptimal, even poor health, with a 50 per cent chance of dying a decade prematurely from homocysteine-related, prevent-able diseases. So, if this is true, who wants to be 'normal'?

Most blood test 'normal' levels have not been based on look-ing at people in optimal health. Most have focused on people who are walking upright, not in the hospital or currently being treated by a doctor and who have no obvious, unusual signs of ill-health. But there's an enormous gap between not being ill, and being positively superhealthy.

For example, going back to coeliac disease: most coeliacs are symptom-free early on. Yet if it's not diagnosed early, cata-strophic medical conditions such as cancer and autoimmune dis-eases can occur later on. In the same way, you may feel fine and have a high H score, but still have all the associated risks of heart disease, strokes and cancer. Clearly, the absence of symptoms and clinical disease does not mean that you are superhealthy.

The trouble is, we tend to think something doesn't exist unless we can 'see' or feel it. That's where high quality medical science comes in. You can't 'see' toxic lead exposure in petrol or a lack of vitamins and minerals lowering children's IQ scores, but it's happening. You can't 'see' or feel your arteries slowly being damaged and blocked. In fact, the first most common symptom of severe coronary artery disease is sudden death! You can't 'see' that a high H score raises your risk of disease, but it does. That's why we say test your H score now and take the necessary preventative action, rather than waiting until diseases can be 'seen'. By then it's often too late. As always, prevention is much better than cure.

The following chapters explain why we are making these recommendations to supplement; in addition they will cover all the issues concerning each of these essential nutrients, such as the best supplement form, safety concerns, and, most importantly, the scientific evidence that they work.

However, if you're convinced already, have been lab tested and know your H score, just choose the supplement combination and dose (see page 185, and see Resources on page 264 for supplement stockists) that best meets your personal needs based on your H score.

■ Supplement with a multivitamin and extra vitamin C

While the above nutrients in the proper amount will successfully lower your homocysteine, we strongly recommend that in addition you take a high-strength multivitamin and multimineral supplement and 1000mg of vitamin C. The reason for this is that there is now unequivocal and substantial evidence that optimising your intake of all essential vitamins and minerals on a regular basis helps to reduce the risk of disease, improve energy, mood, concentration, and resistance to infection, and in short, promote a longer, healthier lifespan. An optimal, high quality multivitamin and multimineral supplement should provide and maintain the levels of folate, B12, B6, B2, magnesium and zinc listed in the chart on page 185, recommended if you are in the 'no risk' category with a homocysteine level below 6.

33

Folate – chances are, you're deficient

WHEN IT COMES to lowering your H score, folate is tops among the B vitamins. But you're very unlikely to be getting enough of it from your diet. Most people's diets provide, on average, 240mcg a day, while the basic minimal requirement is more like 400mcg a day, if not higher. While you will achieve a dietary intake of 400mcg by following the H Factor programme, even this is often not enough to lower a high H score to safe levels. In other words, the higher your H score, the greater your need for folate.

To achieve superhealthy results, the table below shows you how much you need to supplement, in addition to following the diet in Part 4.

Supplement	No risk Below 6	Low risk 6–9	High risk 9–15	Very high risk Above 15
Folate	200mcg	400mcg	1,200mcg	2,000mcg

As the name suggests, folate comes from greens (foliage). The more vegetables and fruit you eat, the more folate you get.

'Whatever your mother or grandmother said about eating spinach – she's right,' says Dr Jacob Selhub of Tufts University in Boston, who has been studying folate for 30 years. He also practises what he preaches, by starting each day with an omelette filled with half a pound of spinach. 'I call folate the vitamin of the next century,' he says. Popeye was clearly on the right track!

There are huge studies on the connection between folate and the prevention of heart disease, strokes, cancer and Alzheimer's disease, to name but a few.[4] The box below shows most of the known diseases associated with folate deficiency. Each disease listed is connected to cell growth. This is because folate is vital to healthy, growing cells: without it, they break down and don't grow or multiply properly. Hundreds of thousands of people in Britain, and over a million Americans, die from these folate-related diseases alone every year.

'If you don't have folate, everything stops,' says Dr JoAnn Manson, an associate professor of medicine at Harvard Medical School in Boston. 'The evidence continues to emerge that folate is beneficial in preventing several important diseases, and I think that folate really does have a staying power that is not true of many of the micronutrients that have been touted in the past.'

Diseases associated with folate deficiency

- Heart attacks and strokes
- Various cancers, e.g. colon cancer and leukaemia
- Diabetes
- Alzheimer's disease
- Neural tube defects in babies, including spina bifida
- Down's syndrome
- Pregnancy problems, including recurring miscarriages and premature births
- Gingivitis – a type of gum disease ▶

- Cervical dysplasia (precancerous condition of the vaginal cervix)
- Anaemia
- Coeliac disease (gluten sensitivity)
- Any disease or disorder of the spinal cord or bone marrow
- Restless leg syndrome
- Disturbances in mood, including depression
- Dementia (brain disorder with diminution of memory, concentration, judgement)
- Schizophrenia-like syndromes
- Insomnia
- Irritability
- Forgetfulness
- Organic psychosis

And, of course, any of the over 50 medical conditions associated with high homocysteine levels in your blood (see Appendix 2, page 237).

Sources: R. Green and J. W. Miller, 'Folate deficiency beyond megaloblastic anemia: hyperhomocysteinemia and other manifestations of dysfunctional folate status', *Seminars in Hematology*, vol. 36 (1999), pp. 47–64; M. Lucock, 'Folate: nutritional biochemistry, molecular biology, and role in disease processes', *Molecular Genetics and Metabolism* (2000), vol. 71, pp. 121–138; G. S. Kelly, 'Folates: supplemental forms and therapeutic applications', *Alternative Medical Review* (1998), vol. 3, no. 3, pp. 208–20.

Of course, you can be folate deficient and not have one of these diseases. The elderly are particularly at risk of deficiency, partly due to poor diet and partly due to poor absorption.[5] A staggering 83 per cent of old people in hospitals are folate deficient. At the other end of the age spectrum, 40 per cent of teenage girls test deficient, along with 53 per cent of otherwise healthy men, aged 50 to 70! As age increases, the need for folate appears to go up. So, how much does an elderly person, or a teenager for that matter, need? Only your H score can tell you for sure.

■ Go green

While the best source of folate is fresh fruit and vegetables, most people just don't eat enough. The average fruit consumption in the UK is a measly *four pieces* a week! In the USA the picture is even worse:

- 85% of all Americans don't meet fruit and vegetable daily recommendations

- 60% of them don't reach the five-a-day bare minimum for fruits and vegetables combined

- 45% have no daily servings of fruit or juice at all, and 22% have no servings of a vegetable on any given day – a certain recipe for folate deficiency and high homocysteine![6]

The ideal intake is at least five servings of a variety of fruits and vegetables a day, which should guarantee you 400mcg. Since folate is a water-soluble B vitamin that only hangs around for a few hours before it's excreted, you need this amount of folate every single day.

■ USA now fortifies food with folate

The US Food and Drug Administration are so concerned about folate deficiency that they have recently passed a law forcing cereal producers to add folate to their products: 140mcg has to be added to every 100g of flour, rice, pasta and cornmeal. They estimate that this will add around 120 to 130mcg of folate to most people's intake. This fortification is not yet happening in Europe but soon will, we predict. In any case, recent research shows that even the FDA recommendation isn't enough if you've got high homocysteine.

■ Pop some folate

Let's assume that you do eat five servings of fruit and vegetables a day. As much as we'd like to say that you can get enough folate from this, scientific research shows clear benefit from supplementing with extra folate.

The effectiveness of supplements over food has also been shown by many researchers. Dr Manuel Malinow and his colleagues at the Providence St Vincent Medical Center in Portland, Oregon gave 75 men and women with heart disease, aged 45 to 85, breakfast cereals fortified with one of three levels of folate. The team then measured the effects of the different diets on the amount of homocysteine in the patients' blood.

Homocysteine levels did not decrease significantly among patients given cereal containing 127 micrograms of folate daily, which is about the amount expected to be gained through food fortification in the USA. H scores did decrease, however, when the patients were given larger supplemental amounts.

This study demonstrates that even a well-balanced diet with folate-fortified foods, although a well-meaning idea, is unlikely to contain enough folate to lower H scores to a safe level and prevent heart attacks. 'We as doctors need to change from saying, "Eat right or take a multivitamin" to saying, "Eat right and take a multivitamin,"' Malinow said.

This is especially important since folate from oral supplements is 1.7 times more effective than folate from food. This means that supplementing 100mcg of folate is equivalent to eating 170mcg in food.[7]

The need for supplements in addition to a folate-rich diet is especially important for those one in ten people born with the gene mutation for MTHFR enzyme.[8] Where this happens, higher levels of folate have been found to help compensate, but dietary sources alone do not supply adequate amounts. In one study, homocysteine levels were much higher (14.5 units) for those with the genetic mutation than for those without it (8.9 units). Although the homocysteine levels of those with the gene mutation remained proportionately higher after upping dietary folate, their homocysteine levels were nonetheless lowered. The addition of a folate supplement had a more marked effect on homocysteine levels than a folate-rich diet alone.[9]

But not only people with the genetic mutation will benefit. It's absolutely vital, for instance, that any woman who is in the first stages of pregnancy, or wishes to become pregnant, supplement daily with at least 400mcg of folate. You don't have to be pregnant, however. Consider this study.

Sixty-five volunteers aged 36 to 71, with an average H score of slightly below 9, were put into three groups. The first group took 437mcg of folate in a daily supplement. The second ate breakfast cereal fortified with 298mcg of folate, which is twice the level recommended by the FDA for fortifying foods. The third group ate folate-rich foods providing 418mg.

The first group who supplemented decreased their H score by 21 per cent. The second group, eating fortified foods, had a 24 per cent decrease, and the third group, eating folate-rich foods, had a mere 9 per cent decrease.[10] Clearly, folate-fortified foods and supplements were the winners here.

Other studies also show that folate supplements or supplemented food work best for lowering homocysteine. At the Northern Ireland Centre for Diet and Health at the University of Ulster and the Departments of Clinical Medicine and Biochemistry at Trinity College, Dublin, people supplemented with 400mcg of folate a day for six weeks. They lowered their homocysteine levels by almost 20 per cent.

However, not everybody responds to supplements of 400mcg alone, which is the level we recommend for those with a low risk. Some studies report lesser H score reductions even at higher daily intakes. For example, one study gave 700mcg of folate in cereal for two months, followed by 2,000mcg of folate in a tablet.

This produced a fourfold increased intake of folate and only an 11 per cent reduction in homocysteine.[11]

Smokers may be particularly at risk. In a study comparing non-smoking Italians with otherwise healthy young smokers, neither group showed any reduction in homocysteine levels after four weeks of taking a 5,000mcg daily folate supplement.[12] The study may have been too short, but it's more likely that in addition to folate, the subjects needed a combination of other homocysteine-lowering nutrients. In another study on 100 Italian smokers who were divided into several groups, homocysteine levels in the group taking 5,000mcg of folate for 45 days were 'significantly reduced'.[13] But once again, taking folate alone, although often effective in lowering high levels of homocysteine, does not work nearly as well as taking it in combination with B2, B6, B2, zinc, and TMG.

In summary

Based on the evidence to date, we recommend that everyone eat a diet high in folate-rich foods. This means at least five servings of a variety of fruits and vegetables each day. In addition to this, we recommend that everybody, even those with H scores below 6, should take a daily multivitamin providing a minimum of 200mcg of folate. Those with an H score of 6 to 9 should supplement 400mcg. Those with an H score above 9 may benefit most from 800mcg, above 12 units from 1,200mcg and above 15 from 2,000mcg a day, until their H score comes down to and is maintained at a safe level.

Folate alone is not enough

However, folate alone doesn't always lower a high homocysteine score and rarely brings it down to a superhealthy level. Other B vitamins, zinc, and methyl donors, especially TMG, are also needed. Of the remaining B vitamins, the second most important is vitamin B12.

34

Vitamin B12 – watch out if you're vegetarian

VITAMIN B12, also known as cobalamin, is the odd one out among the Bs: it's the only B vitamin that isn't found in fruit and vegetables. It is only found in meat, fish, eggs and bacteria-fermented food such as yogurt. It's also found in algae, but this B12 has since been shown to be subtly different from the stuff in meat, and it doesn't correct B12 deficiency.

A study of vegans (who eat no meat, fish, eggs or dairy products) found that pernicious anaemia, the classic disease of B12 deficiency, was much more common in European vegans than vegans in India. Was there something in the Indian diet that contained B12? The answer was 'yes' in the form of an odd insect or two contaminating the Indians' vegan foods! However, vegans eating a more sterilised Western diet run the real risk of being deficient in this vital vitamin.

Whether or not you are vegetarian, we recommend supplementing the following amount of B12, depending on your H score.

Supplement	No risk Below 6	Low risk 6–9	High risk 9–15	Very high risk Above 15
B12	10mcg	500mcg	1,000mcg	1,500mcg

■ Methylcobalamin works best

As far as homocysteine is concerned, the best form of B12 is the methylated, activated form called methylcobalamin. While B12 and folate are both needed to power the enzymes that detoxify homocysteine by adding on a methyl group, methylcobalamin works best in the biochemical context. This is because it is the exact, activated form in which B12 works in the body. Methylcobalamin is also the form most easily absorbed in the intestinal tract and under the tongue. Such 'sublingual' B12 supplements are especially helpful for those with B12 deficiency.

Supplementing extra B12 is doubly important for people at risk of B12 deficiency, and hence high homocysteine levels. See the box below for a list of those with the highest risk.

High-risk groups for B12 deficiency

- The elderly – about 25% are B12 deficient
- Vegetarians and vegans – little or no animal products means little or no B12
- People with gastrointestinal diseases, such as coeliac disease, Crohn's disease, ulcerative colitis, and stomach ulcers and inflammation caused by *Helicobacter pylori* bacteria (these people often have extremely high H scores above 40 units!)
- People on low incomes and undernourishing diets – studies among poorer populations show that 25 to 50% are B12 deficient
- People with autoimmune diseases   ►

- People with AIDs or who are HIV positive
- Those with a family history of pernicious anaemia, which means your body is unable to utilise B12 from the diet.

Source: E. Nexo et al., 'How to diagnose cobalamin deficiency', *Scandinavian Journal of Clinical and Laboratory Investigation Supplement*, vol. 219 (1994), pp. 61–76.

Like folate, B12 is needed for the proper growth and development of all cells, especially nerve cells. This makes it especially important for pregnant women and their babies.

B12 supplements lower homocysteine

While folate supplements do decrease homocysteine on their own, additional supplementation with B12, particularly in the activated methylcobalamin form, can further decrease H scores by as much as an additional 10 per cent. That's what a group of researchers from St James's Hospital, Dublin, found out when they investigated the effect of folate and vitamin B12 on homocysteine status.[14] They also found that 500mcg of folate used in conjunction with B12 worked as well as 5,000mcg of folate, so more is not necessarily better.

As with folate, testing your homocysteine level is one of the most accurate ways of knowing if you are getting enough B12.

B12 is difficult to absorb

How much B12 a person needs is very dependent on how much they can absorb. Unlike other B vitamins, B12 has to combine first with something called 'intrinsic factor', which is produced in the cells that line the stomach. Only then does it become absorbable later on in the digestive tract. Also, there are few absorption sites for B12 in the digestive tract. For these reasons, a person with stomach problems such as ulcers or a lack of stomach acid, or intestinal problems such as coeliac disease caused by gluten sensitivity, can unwittingly become B12 deficient, even when supplementing properly, because their absorption is poor.

Many more people are B12 deficient than is realised. Coeliac disease, for example – notoriously associated with multiple nutritional deficiencies, including deficiencies of folate, B12, B6, magnesium and zinc – goes undiagnosed and untreated in well over 90 per cent of cases. This is serious, as the condition, left untreated, can literally increase your dietary need for B vitamins tenfold. Coeliacs need to be properly diagnosed and then permanently remove all gluten from their diets, after which a vast improvement in B vitamin absorption and lowering of homocysteine naturally happens. See pages 133–5 for more details.

In summary

The best way to know if you are getting enough B12 is to know your H score. If your H score is below 6, then a mere 10mcg of B12, which is what you'll find in a decent multivitamin, will probably suffice. But if you score 6 to 9, we recommend 500mcg. If you score above 9, we recommend 1,000mcg, and above 15 units we recommend 1,500mcg or more. Methylcobalamin is superior, when taken orally or sublingually, to other common forms of B12.

If even these levels of B12, taken in combination with the other homocysteine-lowering nutrients, don't bring down your H score, then we advise twice-weekly intramuscular injections with 1,000mcg B12 (1 millilitre or cc) for two months or so (available from your doctor), since it is possible that you are not producing enough intrinsic factor in the stomach to absorb B12, whatever the dietary or supplemental amount.

35

Vitamins B6 and B2, zinc and magnesium – the forgotten homocysteine busters

LOTS OF US apparently lack some of the key homocysteine-lowering substances – to be precise, vitamin B6 or pyridoxine, vitamin B2 or riboflavin, and the mineral zinc. Although not in the first line of defence against homocysteine, making sure you have enough magnesium also makes a difference. In the USA and UK alike, over half of the population fail to get even the basic requirement of B6 and zinc and are often short in magnesium.

Depending on your H score, we recommend you supplement the following.

Supplement	No risk Below 6	Low risk 6–9	High risk 9–15	Very high risk Above 15
B6	25mg	50mg	75mg	100mg
B2	10mg	15mg	20mg	50mg
Zinc	5mg	10mg	15mg	20mg
Magnesium	100mg	200mg	300mg	400mg

If a supplement contains P-5-P, the activated form of pyridoxine, the B6 amount can be halved because this form is better absorbed and more effective.

Animal protein is particularly rich in vitamin B6, so vegetarians are more likely to be B6 deficient than meat eaters. However, highly processed food, such as fast-food hamburgers, often contain virtually none. So people who live on junk food diets at the local fast food outlet are frequently B6 deficient. Once again, elderly people are often found to be B6 deficient, often due to poor intake and poor protein digestion. The contraceptive pill and oestrogen HRT also deplete B6, increasing some women's need for this vital vitamin. Interestingly, the food colouring tartrazine (E102), found in many processed foods, also depletes vitamin B6 and zinc.

So it's not surprising that a lot of us have symptoms and medical conditions associated with B6 deficiency. While the RDA for B6 is 2mg, most people who are deficient will find that their symptoms and conditions will disappear when they take optimal daily levels of B6, which are around 25 to 100mg.

■ Medical conditions linked to B6 deficiency

Probably the most common symptom of B6 deficiency in women is premenstrual syndrome – nearly half of all women experience it as bloating, carbohydrate cravings, depression, irritability, cramping and fluid retention. Taking 100mg of B6 has proven to help relieve these symptoms.

B6 is vital not only for digesting protein, but also for almost all stages of protein chemistry in the body. Homocysteine is a case in point, being derived from the protein constituent methionine. Protein has to be well controlled in the body, otherwise the acid/alkaline balance goes out of whack and serious conditions can ensue. Since excess protein is 'buffered' by calcium, for instance, eating too much protein can lead to osteoporosis. And in susceptible people, recurring calcium oxalate kidney stones, by far the most common form of kidney stones found in people in the US and UK, are often produced by a deficiency of B6 and/or magnesium. Women who supplement more than 40mg of B6 a day get fewer kidney stones, while supplementing larger amounts, such as 250mg a day, has been shown to prevent and reverse stones in some sufferers.

Low B6 is also a risk factor for high blood pressure, independent of homocysteine. Scientists believe the reason is that when B6 is low, there are increased levels of inflammation; this is shown by an increase in the blood protein C-reactive protein, an excellent marker for inflammatory diseases such as heart disease.

In a study of 891 Americans, participants were divided into two groups. The first group had lower levels of C-reactive protein in their blood (below 6mg/L) and the second higher levels (above 6mg/L). When researchers tested for vitamin B6, they found that the second group with high C-reactive protein had 30 per cent lower B6 levels.[15] Another more recent study showed a strong relationship between high levels of C-reactive protein and homocysteine. With C-reactive protein pretty much accepted now as a marker for inflammation, it is often high, accompanied by high homocysteine levels and low B6, in other inflammatory diseases, including rheumatoid arthritis, diabetes, heart disease and certain cancers.

The table below lists many of the other conditions and symptoms associated with B6 deficiency.

Conditions linked to B6 deficiency

- PMS
- Depression
- Autism ▶

- Carpal tunnel syndrome
- Kidney stones (calcium oxalate)
- Asthma
- Heart disease
- Hypertension
- Hip fractures
- Rheumatoid arthritis

Symptoms of B6 deficiency

- Fatigue
- Mood swings
- Irritability, agitation
- Insomnia
- Loss of appetite control
- Decreased tolerance to pain
- Poor coordination
- Difficulty breathing
- Weight gain, weight loss
- Fluid retention
- Epilepsy in children
- Morning sickness

How B6 and zinc lower homocysteine

Unlike B12 and folate, which help turn homocysteine into the body's best methyl donor SAMe, B6 is vital for turning homocysteine into glutathione, the body's best antioxidant and detoxifier (see Figure 3 on page 25). The two key enzymes that make this vital conversion (cystathionine beta-synthase and cystathionine lyase) both depend on vitamin B6, as well as B2 and zinc.

This means that when you are low in folate or B12, or have the

MTHFR gene mutation, your homocysteine levels will zoom up, and you'll also need more B6, zinc and B2 to lower them.

Zinc, found in fish, poultry, meat, egg yolk and seafood, and also in nuts, seeds and wheatgerm, is the most commonly deficient mineral. Vegetarians may need as much as 50 per cent more zinc than non-vegetarians because of the lower absorption of zinc from cereals and other plant foods, so it is very important for vegetarians to include good sources of dietary zinc. Up to 50 per cent of all alcoholics are deficient in zinc due to alcohol inhibiting absorption and increasing urinary loss of zinc. Diabetics and people with chronic kidney disease need more zinc. Also, people suffering from chronic diarrhoea, coeliac disease or Crohn's disease are also commonly zinc deficient (notice that all these conditions listed are at an increased risk when high homocysteine levels are present).

Zinc is especially vital in the 'homocysteine dance' because B6 can't work without it. Before B6, or pyridoxine, can activate the enzymes that clear homocysteine, it must first be converted into pyridoxal-5-phosphate, also called P-5-P, which is the 'activated' form of B6. This conversion is completely dependent on zinc. Not surprisingly, when someone has a low level of pyridoxal-5-phosphate (P-5-P) in the blood, they are often high in homocysteine,[16, 17, 18, 19] even when folate and B12 levels are adequate.

If you are deficient in B6, supplementing with just 1.6mg of zinc can lower homocysteine levels by nearly 8 per cent. This remarkable finding came from researchers from the Northern Ireland Centre for Diet and Health at the University of Ulster and the Departments of Clinical Medicine and Biochemistry at Trinity College, Dublin. They gave 22 healthy elderly people this low daily dose of B6 for 12 weeks (after first ensuring supplementation to achieve adequate levels of folate and B12) and found that homocysteine levels were reduced by another 7.5 per cent.

Supplementing B6 at 20mg a day for six weeks significantly decreased homocysteine levels in those who were B6 deficient, but was not sufficient to decrease levels in those who weren't.[20] Sometimes, negligible results with B6 supplementation happen because the person doesn't need more B6, or because the amount given isn't high enough, or because the person is low in zinc and/or B2, and is therefore unable to activate the B6. The average daily zinc intake in the US and Britain, remember, is 7.6mg or less which is less than half the RDA of 15mg. Also, the amount of cereal grains we eat can affect our absorption of zinc. As a result, zinc deficiency in the UK and US is extremely common.

For this reason we recommend supplementing with both vitamin B6 and zinc. Even better would be to provide some P-5-P, the activated form of B6, when you do.

For those adults with H scores below 6 we still recommend supplementing a basic level of B6 (25mg) and zinc (5mg) every day, which is what you can find in a high quality multivitamin/multimineral. If your H score is 6 to 9, we recommend 50mg of B6 and 10mg of zinc. If above 12 units, 75mg of B6 and 15mg of zinc, and if above 15 units a maximum of 100mg of B6 and 20mg of zinc.

■ Magnesium

Magnesium is in the second line of defence against homocysteine. In order to convert homocysteine to SAMe by way of methionine, magnesium is required. If magnesium is deficient, high homocysteine, low SAMe and poor methylation will follow. Unfortunately, magnesium deficiency due to poor diets, soft water and overuse of diuretics (water pills) is extremely common, and is associated with many medical conditions.

There are at least 30 well-defined medical conditions associated with magnesium deficiency. And predictably high homocysteine shares many of these conditions with low magnesium. These include:

- Atherosclerosis

- Coronary heart disease, heart attacks

- Coronary bypass surgery

- Strokes, transient ischemic attacks

- Migraine headaches

- Major depression

- Chronic fatigue

- High blood pressure (at least 20 per cent of the elderly have magnesium deficiency)

- Type II diabetes (40 per cent or more have low magnesium levels)

- Abnormal blood clotting (thrombosis)

- Alcoholism

- Coeliac disease

- Crohn's disease

- Pregnancy problems, including pre-eclampsia

- Cancer.

Due to the very close relationship between high homocysteine, low magnesium and the above medical conditions we strongly recommend that magnesium be included in the H Factor programme. Fish, brown rice, fruits, spinach and other vegetables, nuts and seeds are loaded with magnesium. Most good multivitamins should provide 200mg of magnesium, but many don't. If you are supplementing a good quality multivitamin, and eating these foods, you probably don't need extra magnesium. The best absorbed oral magnesium supplement appears to be one that is bound to an amino acid. This is known as chelated magnesium. Also good is magnesium citrate. One of the better chelated magnesium supplements is magnesium glycinate (magnesium attached to the amino acid glycine). Glycine gives you an extra bonus by increasing glutathione levels.

If your H score is less than 6 units, we recommend 100mg of magnesium daily. For an H score between 6 and 9, 200mg. If your score is anywhere between 9 and 15, 300mg. And for a very high H score over 15 units, we recommend 400mg of magnesium daily.

■ Vitamin B2 (riboflavin)

As you can see from Figure 3 on page 25, vitamin B2 (riboflavin) helps activate B6 and drive the enzymes that turn homocysteine into both glutathione and SAMe. Making sure you have enough B2 is therefore important, especially for those with the defective MTHFR enzyme.

When this enzyme is impaired, invariably leaving homocysteine levels high, scientists have found that vitamin B2 insufficiency is also common. In a study of 286 volunteers, those with the MTHFR mutation had the lowest B2 levels and an extremely high average homocysteine score of 18 units – almost twice that of the group without the genetic mutation. However, those who had adequate B2 in the enzymatically impaired group had a homocysteine score comparable to that of the non-MTHFR-defective group.[21]

In a further study designed to investigate the relationship between B2 and homocysteine, scientists examined 423 healthy blood donors aged 19 to 69 years. They divided people into four groups, from the 25 per cent with the lowest B2 levels up to the 25 per cent with the highest B2 levels. They found that homocysteine was 1.4 units higher in the bottom 25 per cent, based on

B2 status, compared to those in the top 25 per cent. But further investigation showed that the people with higher homocysteine levels also had a defective methylation enzyme (MTHFR) gene.[22]

Therefore, we recommend that adults with a low H score supplement daily with 10mg of vitamin B2, which is what a good multivitamin provides. If your H score is 6 to 9, we recommend 15mg of B2. If 9 to 12, 20mg. If above 12, 30mg of B2, and if above 15 units, 50mg of B2 every day.

36

SAMe, TMG and choline – the premier methyl donors

THERE ARE TWO BASIC, indispensable elements in the task of keeping your H score low. One is the methyl movers – that's folate, B12, B6, B2, magnesium and zinc, all of which keep your body's chemistry in balance by helping homocysteine-lowering enzymes to work. The other is simply having enough methyl groups in the first place. It's these methyl groups that the enzymes stick on to homocysteine to turn it into something that's good for you, not bad.

A nutrient that 'donates' methyl groups is called a methyl donor. There are three main players here – choline, TMG and SAMe, all of which help lower homocysteine, and are available in supplements. Of these three, TMG is the most powerful, and provided you are supplementing it, you don't need the other two to help bring down your H score. Here's how much you need to supplement with TMG.

Supplement	No risk Below 6	Low risk 6–9	High risk 09–15	Very high risk Above 15
TMG	500mg	750mg	1.5–3g	3–6g

These nutrients need a good supply of methyl groups to get rid of toxic homocysteine. And the higher your H score, the more methyl groups you need to bring it down. As you can see from the figure below, choline has four methyl groups, TMG has three and DMG (dimethylglycine) has two. However, DMG isn't commonly used to lower homocysteine. The same is true for choline which must first be converted into TMG; in other words choline itself doesn't directly lower homocysteine, but TMG does. This is why we recommend TMG as the best homocysteine-lowering nutrient.

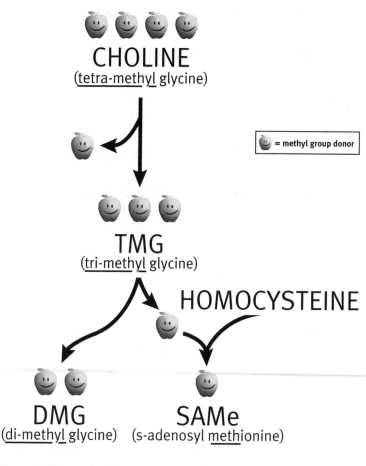

Figure 7. The methyl donors

Normally, methyl groups are supplied by TMG and SAMe. The body takes a methyl group from TMG, using B12 and folate as 'methyl magnets'. This turns them into methylated (activated or supercharged) forms of the vitamins, called methyl-folate and methylcobalamin (B12). They then detoxify the homocysteine by giving up their methyl groups and, in so doing, helping convert homocysteine back to methionine which, with the help of magnesium, is in turn converted to SAMe.

■ Best foods for methyl donation

There are some foods rich in choline and TMG. Choline, for instance, is found in egg yolk and organ meat. TMG, sometimes called betaine, is found in sugar beet, beans, peas, prawns, fish, eggs and especially liver. SAMe, as we've seen, isn't found in food, but rather made in the body solely from methionine by way of homocysteine methylation; that is, provided you've got enough B vitamins and magnesium. Of all these methyl donors, TMG is the most important in your diet and SAMe is by far the most important in your body. This is because your body can most effectively make more SAMe, as well as dramatically reduce homocysteine, with the help of TMG. Assuming you're well nourished with B vitamins and magnesium, supplementing with 3g of TMG effectively makes 1g of SAMe in the body.

■ SAMe – the master tuner

We've seen how SAMe is made in the body, and how having enough vitamin B6, B12, folate, B2, magnesium and TMG is essential in that process. And the lower your homocysteine, the higher your SAMe levels, so by lowering your H score you effectively raise SAMe levels in the body and brain.

Now for a closer look at SAMe, the body's biochemical master tuner.

In your brain, SAMe helps make or activate three highly important neurotransmitters – serotonin, melatonin and dimethyl tryptamine.

Serotonin keeps you happy and less irritable, helps alleviate chronic pain, and cuts down on carbohydrate/sweet cravings. A deficiency in serotonin leads to a lowered mood and depression, and that's why over 100 medical studies have proven SAMe to be a highly effective anti-depressant.[23] One meta-analysis of all the studies on SAMe found that an impressive 92 per cent of depressed patients responded well to SAMe supplementation, compared to 85 per cent for standard anti-depressant medications. SAMe also works fast. 'Most people taking SAMe see some effect within 10 days,' says Dr Teodoro Bottiglieri, director of neuropharmacology at Baylor University Medical Center in Dallas, Texas, one of the top experts on SAMe. SAMe is also highly touted as a proven therapy for the pain and stiffness of osteoarthritis and as a protector against liver disease.

Melatonin is activated by SAMe. Although it is a hormone, melatonin is also considered to be a powerful antioxidant. It also helps to generate that other powerful antioxidant, glutathione, and by doing so helps keep potentially toxic oxygen radicals from building up and harming brain cells, the liver and your DNA. Melatonin is thought to be protective against Alzheimer's disease, to improve short-term memory and to protect against certain kinds of cancers (for example, some studies indicate it may reverse cancer of the prostate!). When melatonin-deficient, you suffer from poor sleep quality, depression, chronic fatigue, loss of memory, and irritability.[24]

Melatonin is perhaps best known for keeping you 'in sync' with nature and the rhythm of day and night and the seasons. Produced in the pineal gland, the 'third eye' in the centre of the head, melatonin levels rise when you sleep, and help to promote sleep and dreaming. All the American astronauts when in orbit take melatonin to sleep, dream and keep their circadian rhythms in check.

And what of dimethyl tryptamine? This is the mind's own consciousness-expanding neurotransmitter, keeping you 'connected' to the big picture of life. So SAMe, by helping make all these fascinating brain chemicals, is much more than an antidepressant. It also promotes a sense of wellbeing, and helps you stay calm and connected.

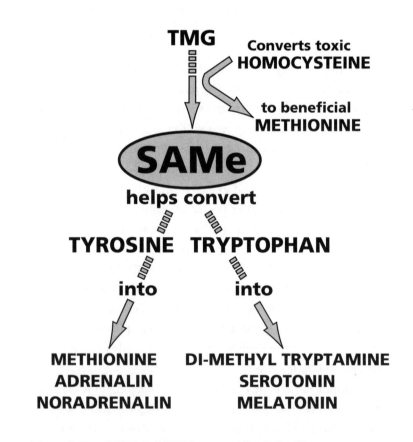

Figure 8. How SAMe and TMG keep your brain healthy

But these aren't the only neurotransmitters SAMe is involved in manufacturing. As you can see from the figure above, the motivating neurotransmitters dopamine, adrenalin and noradrenalin are made by SAMe, too. Your levels of noradrenalin are high when you're in love. Adrenalin is high when you're ready for action. And almost every pleasure state of the body, from having sex to eating chocolate, involves dopamine. So SAMe acts as a natural stimulant and motivator, helping keep you on an 'up'. Most people experience these benefits within days of supplementing SAMe, although like most supplements the best effects are experienced after weeks of supplementation.

In the brain, this happens in several ways. Brain nerve cells, called neurons, are built out of special fats called phospholipids. But first these have to be methylated (sound familiar?). Enter your primary intracellular methylator, SAMe. SAMe literally helps build your entire brain and nervous system, improving the body's ability to communicate.

And it isn't just your brain that SAMe helps keep connected. It's also your body. By methylating, SAMe helps to build and rebuild connective tissue, which is literally the glue between cells that helps to keep you together. This means healthier joints, firmer skin and fewer wrinkles! Not only does SAMe help relieve the pain and stiffness of arthritis, it actually helps rebuild cartilage. It also helps reduce the inflammation and pain associated with fibromyalgia, an increasingly common disease especially in women involving generalised muscle pain and localised tenderness in certain areas of the body. SAMe also plays a key role in protecting your liver from injury or disease and stimulates enzymes that produce glutathione, our body's primary antioxidant and detoxifier.

So remember, as you lower your homocysteine level you are maximising the brain and body's availability of SAMe, the master tuner.

Supplementing SAMe

With all these benefits you might be wondering why we don't all take SAMe. In a very real sense we can, simply by lowering homocysteine with the use of B vitamins, magnesium and TMG.

You can, however, supplement with SAMe, although as we warned you in Part 1, it's a relatively expensive option. If you choose to supplement with SAMe, it's best taken on an empty stomach, preferably 20 minutes or more before, or one hour after, food. Since some people initially complain of mild digestive upset with the higher doses of SAMe, we recommend that you start with a dosage of 200mg twice daily. If you don't see results in a few days, you can gradually increase it up to a maximum of 400mg four times daily, if needed. Most often, 400mg per day is sufficient. Remember that SAMe should also be taken with its cofactor, magnesium, and homocysteine-lowering vitamins B6, B12, folate and B2. SAMe appears to be safe to use during pregnancy and breastfeeding.

SAMe can't be supplemented in its pure form: it is too unstable. So it comes in supplements bound to one of two stabilising compounds, tosylate and butanedisulfonate. The former quickly degrades upon exposure to heat and/or moisture. However, both forms are relatively stable in enteric-coated tablets, which means that they aren't destroyed by stomach acid, thereby helping to stabilise and protect the SAMe. SAMe can then be safely released lower in the digestive tract, where it can be absorbed in an active form. The two stabilising compounds also reduce the chances of nausea and gastrointestinal disturbances sometimes reported with SAMe supplementation. Reducing the SAMe dose size and taking it with meals usually overcomes this potential problem, although taking it with meals may reduce its potency somewhat.

So it's best to buy stabilised SAMe supplements containing either the tosylate or butanedisulfonate form. Bear in mind that you want to know how much 'active' SAMe there is, not just the amount of the total compound, since tosylate and butanedisulfonate are heavy compounds, adding to the total weight of the tablet. For example, 200mg of SAMe with butanedisulfonate provides only 100mg of SAMe. Another caveat: purchase from an established reputable company, not newly on the SAMe bandwagon, and be prepared to pay for the quality.

Or to reiterate, you can choose to make your own SAMe. By lowering homocysteine to a safe level by following our H Factor programme, you will have significantly increased your own production of SAMe. For example, it's been shown that when folate levels go up, so too does SAMe, and homocysteine levels go down.[25]

▪ Trimethylglycine (TMG)

While there's no question that SAMe supplements work in lowering homocysteine, there are some major advantages to increasing your intake of TMG instead, and letting your body make its own SAMe. One advantage is that TMG, in the process of helping generate more SAMe, also detoxifies homocysteine by methylating it, in the process lowering your disease risk, thus killing two birds with one stone.

TMG lowers your H score

Few researchers have acknowledged the importance of supplementing TMG to lower homocysteine. Most have instead latched on to folate and B12 supplementation, ignoring the boatload of evidence that a full combination of homocysteine-lowering nutrients work much better than any one or two in isolation. We'll explore this phenomenon in detail in the next chapter.

Supplementing TMG in addition to B6 and folate can help to lower homocysteine levels. In a study in New Zealand, the H score of patients with chronic kidney failure and very high homocysteine levels was reduced by a further 18 per cent when 4g of TMG was given, along with 50mg of B6 and 5,000mcg of folate, compared to patients taking just B6 and folate.[26]

In another study, 15 'normal, healthy' people, aged between 18 and 36, were given 3g of TMG alone, twice a day. At the start of the trial, the average homocysteine level was a moderately elevated 10.9 units. After three weeks, the homocysteine level was reduced by over 13 per cent. The researchers concluded that 6g of TMG daily does in fact lower homocysteine, even in so-called healthy people with 'normal' homocysteine levels (as we have earlier pointed out, 10.9 units is the average, normal level in the US and UK, but is definitely an unhealthy level). But they noted that the extent of the decrease is much smaller in 'healthy' volunteers than in high-risk patients with much higher homocysteine. For example, in patients with H scores above 50 units, TMG supplementation can lower the scores by up to 75 per cent.[27]

On its own, TMG supplementation tends not to be as good as folate. However, researchers from the Netherlands have discovered that folate *and* TMG may work much better as partners because TMG encourages homocysteine detoxification mostly in the liver, whereas folate supports this process in cells all over the body. To repeat, TMG works best when taken in conjunction with all the other homocysteine-lowering B vitamins.

Boosting levels of TMG may be particularly beneficial if you're overdoing alcohol and your liver is under a lot of strain. TMG certainly works for rats. Like humans, rats fed high-alcohol diets develop high levels of homocysteine, now thought to be one of the ways that alcohol damages the liver. TMG helps to lower toxic levels of homocysteine, thereby protecting the liver from alcohol-provoked injury.[28]

It is even possible that the presence of TMG (betaine) in French wine might play an important part in the low level of heart disease in France. France has one of the lowest rates of heart disease in Europe, despite a high consumption of wine, cholesterol and foods rich in saturated fat – a phenomenon called the 'French Paradox'. Researchers from the School of Public Health at the University of North Carolina believe a possible reason for this could be lower homocysteine levels as a result of drinking red wine. This is because TMG is often added, as betaine, to less expensive wines in beet sugar used to increase alcohol content. The team found that an average glass of wine contains approximately 3mg of TMG.[29] Is this enough to significantly lower homocysteine?

TMG – should we all supplement and if so, how much?

Hopefully, we've convinced you by now that it's vital to have an optimal supply of methyl groups at all times throughout the body and brain, controlling how we think and feel, protecting us from heart attacks, strokes, cancers, liver and autoimmune diseases as well as slowing down the ageing process. This is a compelling argument in favour of anyone, even those with an H score under 6 units, to supplement daily with TMG.

If your H score is under 6 units, we recommend 500mg of TMG daily. Between 6 to 9, we recommend 750mg. Between 9 and 12, 1,000mg. Between 12 and 15, 1,500 to 3,000mg, and if above 15 units, 3,000 to 6,000mg a day, split into 1,500 to 3,000mg twice a day, preferably on an empty stomach.

Assuming that you're well nourished with B vitamins and magnesium, with TMG's help your body will make enough SAMe, in a roughly three-to-one ratio; in other words, 300mg of supplemental TMG is equivalent to 100mg of SAMe.

■ Choline

TMG is the single best dietary or supplement methyl donor. TMG is naturally produced in your body from choline. So in addition to supplementing with choline, there are good reasons to make sure you are taking in enough choline from eggs, liver and other organ meats, preferably organic. Good to excellent sources of choline are listed below:

Choline content of common foods

Food/serving	Mg of choline/serving
Beef liver, 85g (3oz)	453.2
Egg, 61g (2oz or 1 large)	345.0
Beef steak, 85g (3oz)	58.5
Cauliflower, 99g (3oz or ⅙ medium head)	43.9
Iceberg lettuce, 89g (3oz or ⅙ medium head)	28.9
Peanuts, 28g (1oz)	28.3
Peanut butter, 32g (1oz or 2 tbsp)	26.1
Grape juice, 236ml (8oz)	12.9
Potato, 148g (5oz or 1 medium)	12.9
Orange, 154g (5oz or 1 medium)	11.5

Much of the choline in the body gets used to make phosphatidyl choline, a type of phospholipid from which the body can build new nerves and brain cells. Phosphatidyl choline is also the starting point for making the brain's memory molecule, a neurotransmitter called acetylcholine. Low levels of acetylcholine are found both in children with learning difficulties and in older people with memory decline.

One of the easiest ways to up your intake of choline is to sprinkle some lecithin granules on your cereal in the morning. Lecithin is exceptionally high in phosphatidyl choline. A dessertspoon a day not only helps give the body TMG-donated methyl groups, it also helps keep your memory intact.

Caveat emptor: not all lecithin is equal. Most lecithin is around 15 per cent phosphatidyl choline. Some commercial lecithin products are much more highly concentrated, providing 30 per cent phosphatidyl choline. An ideal daily supplemental intake of choline is perhaps 1,200mg. That's a dessertspoon of regular lecithin, or a heaped teaspoon of a highly concentrated (30 per cent) phosphatidyl choline lecithin. Alternatively, there are brain-friendly supplements that provide both phosphatidyl choline and phosphatidyl serine, another memory-boosting nutrient.

By ensuring an optimal intake of these phospholipids, you are protecting both your brain and body, and helping to keep yourself well methylated, which, in turn, helps to lower your homocysteine level.

See Appendix 1, page 236, to see how the body and brain keep a perfect balance of choline, phosphatidyl choline and TMG.

37

The synergy principle – why multiple nutrients work better

NUTRIENTS LOVE COMPANY. So while there is plenty of evidence that each of the key nutrients discussed in the last four chapters – folate, B12, B6, B2, zinc, and TMG – work on their own to lower homocysteine, they're much, much better together: their impact is greater than the sum of all the individual effects. In other words, one plus one equals three – or four! This is the principle of synergy.

The idea that nutrients work better in combination is not only common sense; it is also proven. First of all, people rarely, if ever, become deficient in just one essential nutrient. A person eating fewer fruits and vegetables, nuts or seeds is going to be deficient in a wide range of nutrients. Hence, we recommend eating a highly varied and wholefood (unprocessed, unpackaged, uncanned) diet, giving you the best full spectrum of nutrients. We also passionately recommend taking a high quality, comprehensive multivitamin-multimineral supplement that provides optimal amounts of all the B vitamins, and zinc.

We've discussed how the enzymes involved in turning homocysteine either into SAMe and glutathione depend on quite a crew: B12, B6, B2 and zinc, plus having enough methyl groups available, generously donated by TMG. So you'll need the right combination of synergistic (one plus one equals three) nutrients. The boost in effectiveness given by synergy also means lower amounts of each key nutrient is needed, at less cost to you.

This is exactly what scientific research and published studies show. While conventional medicine likes to focus on one 'drug' for one condition (folate to lower homocysteine, for example), 20 per cent of people with high homocysteine don't respond to folate on its own. Paradoxically, there's at least one study showing people taking only folate have a modest *increase* in their H score!

This was demonstrated in a published study of 304 men and women who were given either 1,000mcg or 2,000mcg of folate for three consecutive weeks. While overall there was a small decrease in average H scores, 20 per cent of people in the study had a modest increase of 1.5 units. No obvious reason for this individual variation was given.[30] So folate on its own is sometimes ineffective, but not so in combination with other homocysteine-lowering nutrients.

The power of synergy is well illustrated again by a very important study from Japan involving patients with chronic kidney disease, a condition that is linked with very high homocysteine levels, and hence heart attacks and strokes.

The Japanese patients were divided into three groups: one was given either folate or B12 alone (in the form of methylcobalamin, the active form of B12), one group folate and B12 together, and the third group folate, B12 and B6. The trial lasted for three weeks.[31]

Here are the results of this remarkable study:

Supplement group	Homocysteine change
Folate alone	17.3% reduction
B12 alone	18.7% reduction
Folate plus B12	57.4% reduction
Folate, B12 and B6	59.9% reduction

Notice that this extraordinary study has revealed two very important principles. First, the more nutrients provided, the greater the reduction in homocysteine. And secondly, the right combination of nutrients at the right dose can more than halve your H score in as little as three weeks!

This study shows that adding in B6 provides only a small additional benefit. However, we question the lack of zinc and B2, all of which are needed to activate B6. An analysis of 12 other trials, involving a total of 1,114 people, came to a similar conclusion.

The researchers found that folate on its own could lower homocysteine by 25 per cent. Adding in B12 produced a further 7 per cent reduction, giving a 32 per cent reduction overall. On the other hand, they did not find a significant further reduction with the addition of very low doses of B6 (averaging 16.5mg).[32] While B6 may not always be as powerful a homocysteine buster as folate or B12, we think that larger amounts of B6 (including activated B6 as P-5-P), together with zinc and B2, would be more effective. Then, of course, last but not least, TMG should always be included because it provides the much-needed methyl groups to help detoxify and lower homocysteine.

These combined nutrients not only dramatically lower your H score, but more importantly, reduce your risk of disease. A recent trial, published in the *Journal of the American Medical Association,* gave patients with a very high risk of heart disease and high homocysteine either a combination of folate (1,000mcg), B12 (400mcg) and B6 (10mg), or a placebo. Within one year, 15.4 per cent of the patients taking B vitamins experienced a major coronary event, while 22.8 per cent of the patients in the placebo group experienced major adverse events. The supplemented group had reduced their risk of death from heart disease and heart attacks[33] by one-third in only one year! One is forced to wonder how much further the risk would have been reduced had they also added B2, zinc and TMG – and perhaps magnesium, too.

Because of all this, you'd expect to find that people who take multivitamins – which generally provide folate, B12, B6, B2 and zinc – would have lower H scores and a lower risk of disease than those who don't. Exactly such a comparison has been made between 256 multivitamin supplement users versus 230 non-users. Remarkably, supplement takers had one-half the H score of the others! This study also showed that eating enough protein and lots of fruit and vegetables, and not smoking, were all associated with having a lower H score.[34]

No one to our knowledge has yet done the 'gold standard' test by simultaneously giving those high-risk individuals with high homocysteine all the methyl movers (folate, B12, B6, B2 and zinc), plus the mother of all methyl donors, TMG. But this is in fact the strategy we have found most effective in lowering your H score quickly.

38

Supplements – a word of caution

IN AN AVERAGE YEAR, 2 million Americans in hospitals have serious adverse drug reactions, and 180,000 die. This makes deaths from prescribed drugs the fifth largest cause of death in the US, according to the *Journal of the American Medical Association*. Although no similar study has been done in Britain, we are confident the findings would be very similar: adverse drug reactions, including the fatal ones, are no less common in the UK.

This is clearly a compelling argument for alternative medicine. But what about bad reactions to natural medicine and nutritional supplements? A few dozen people have allegedly died from herbs (ephedra/ma huang being one example), but to our knowledge no one has ever died after taking a B vitamin, zinc, magnesium or TMG supplement. Even minor adverse reactions are extraordinarily rare. This illustrates how nutrients have a substantially larger margin of safety built in.

However, this does not mean that more is necessarily always better. Too much of a good thing can be bad. While we are confident that you will encounter no significant toxic or adverse effects with even the highest amounts of nutrients we've recommended, we'd like you to be fully informed about the potential hazards of some supplements when taken in excess.

▪ Vitamin B6 and zinc toxicity

Vitamin B6, in doses of 500mg or more, taken for months, not weeks, can cause 'peripheral neuropathy', a type of nerve damage with symptoms including tingling and numbness of the hands or feet. There is even a study or two suggesting that toxicity may occur with 150mg B6 for prolonged periods. This neuropathy is almost always reversible. However, to be on the side of caution, we do not recommend you take more than 100mg of B6 a day, which is completely safe for long-term use.

Zinc antagonises copper, and vice versa. This means that if you are copper deficient and supplement more than 50mg of zinc for three months or longer, you may become more copper deficient. Almost all whole foods contain copper – including beans, lentils, nuts, seeds, unrefined flour, rice and so on. Provided you eat some of these foods on a regular basis, you are unlikely to become copper deficient when supplementing with zinc. On this basis, 20mg of zinc a day, the highest dose of zinc we are recommending, is completely safe for long-term use. However, to be extra cautious, we do not recommend that you take more than 20mg of zinc for longer than three months. In addition, large doses of zinc, especially on an empty stomach, can cause slight reversible nausea in some sensitive people. If this happens, we recommend you take your zinc-containing supplement with food, even though you will absorb slightly less.

▪ Magnesium

Too much magnesium may cause diarrhoea, easily reversed by reducing the daily dose. With the sole exception of a retarded infant who reportedly died as a result of being given enormous daily doses of magnesium for constipation (2,400mg a day, three times the maximum recommended daily dose for infants, and six times our highest recommended dose for adults with very high homocysteine), reports of magnesium overdose and toxicity are essentially non-existent.

■ Vitamin B12 and folate

Neither of these vitamins is associated with adverse effects, even at very large amounts above those we've recommended. However, giving supplements of folate alone to a person who is B12 deficient can mask the symptoms of pernicious anaemia, the first obvious symptom of which is usually chronic fatigue. In such a case the underlying condition gets worse, even though energy improves. For this reason we do not recommend ever supplementing folate on its own, but rather recommend always supplementing it in conjunction with vitamin B12. Multivitamins invariably contain both B12 and folate.

■ SAMe and TMG

In some people high doses of SAMe, above 1,000mg a day, have been known to induce nausea. It is therefore better to start with a lower dose, such as 200mg twice daily and work up after several days. If you do get nauseous with SAMe, either switch to TMG (along with the other homocysteine-busting nutrients) or take the SAMe with food. While this does reduce the likelihood of nausea, it also reduces the overall homocysteine-lowering effect.

Large amounts of TMG, usually above 2,000mg, very infrequently cause headaches in some people. Above 5,000mg a day is associated with a sulphurous body odour. If this happens to you, lower the amount of TMG taken at any time by spreading doses throughout the day.

■ Caveat emptor: high doses of vitamins B3 and C can raise homocysteine

Niacin, also called vitamin B3, the 'flushing B3' or nicotinic acid, is needed to make energy in the body. So if you're not getting enough of it, all sorts of things go wrong. You feel tired, can't think straight, your skin starts to get all itchy, your bad LDL cholesterol goes up, your HDL cholesterol goes down and you get diarrhoea. The basic RDA for niacin is 18mg a day. Most decent – and science-based – multivitamins give you about 50mg a day of niacin or niacinamide.

Niacin has been used effectively to reduce elevated LDL cholesterol levels while elevating the good HDL cholesterol – but when supplemented alone in very large amounts, it may not be such good news for homocysteine. In a trial of 52 patients with peripheral vascular disease, half were given a placebo each day and the other half received up to 3,000mg of niacin – almost 200 times the RDA. After 18 weeks, average homocysteine levels in the niacin group had *increased* a whopping 55 per cent, from 13.1 units to 21.1 units![35] This effect has not been shown below 1,000mg of niacin, which we think is a safe upper limit.

The likely reason these very high amounts of niacin raise homocysteine is that the removal of niacin from the body requires it to become methylated. In a person with high homocysteine due to lack of enough methyl groups, this robs vital methyl groups needed to detoxify homocysteine, thus dangerously raising H scores. The solution to this undesirable side-effect is to lower niacin supplements a bit and always take the premier methyl donor, TMG, with niacin.[36]

So, since large amounts of niacin can be highly effective in

lowering bad LDL cholesterol while raising good HDL cholesterol, we recommend that anyone taking more than 500mg of
niacin daily also take equal amounts of the methyl donor TMG.

500mg vitamin C

For reasons as yet unexplained, vitamin C supplements, when
used alone in doses of 500mg daily, slightly raise homocysteine.
This was reported in an Italian trial in which people with high H
scores were given either folate, which predictably lowered H
scores, or 500mg of vitamin C. Unexpectedly, there was a slight,
although statistically not significant, increase in homocysteine in
those taking the vitamin C.[37] Until other studies confirm or
refute this finding, as a precautionary step we recommend that
anyone supplementing with vitamin C at or above 500mg daily
also take a multivitamin providing reasonable amounts of the
homocysteine-lowering B vitamins – and periodically monitor
your homocysteine blood level just to be safe.

Ten Steps to Superhealth

Ten steps to superhealth

PREVENTION IS NOT ONLY better than cure; it is guaranteed to save and extend your life. Most of the diseases all of us want to avoid – heart attacks, strokes, cancer and Alzheimer's – don't have any guaranteed cures. But they can be prevented.

Knowing your homocysteine level and taking these ten simple steps to bring your H score down to 6 or below is the most concrete way of minimising your chances of ever having a preventable, life-threatening disease. It is one of the best presents you could ever give yourself. The effort and costs are minimal, the gain is likely to be additional years of healthy living, with immediate side benefits in terms of increased energy and wellbeing.

Here's all you have to do.

1 Begin by testing your homocysteine level.

2 Follow the ten-step H Factor diet, below.

3 Take the H Factor supplements at the dose prescribed below, depending on your H score.

4 Retest after three months and adjust the supplements accordingly – and when you've achieved a superhealthy level, retest every six to 12 months to ensure you're maintaining it.

Our recipe for a long and healthy life is to keep doing this until your H score is no more than 6, then keep following the H Factor programme to keep your health high and your homocysteine low.

The ten-step H factor diet

■ Eat less fatty meat, more fish and vegetable protein

Eat no more than four servings of lean meat a week; unfried, unbreaded, unsmoked fish at least three times a week; and if you're not allergic or intolerant, a serving of a soya-based food (such as tofu, tempeh, soya milk or soya sausages) or beans, such as kidney beans, chickpea hummus or baked beans, at least five times a week.

■ Eat your greens

Have at least five servings of fruits or vegetables a day. This means two pieces of fruit every single day, and three servings of vegetables. Vary your selections from day to day. Make sure half of what's on your plate for each lunch and dinner is vegetables. A bountiful mixed salad each day with extra virgin olive oil also makes excellent sense.

■ Have a clove of garlic a day

Either eat a clove of garlic a day, or take a garlic supplement every day. You can either take garlic oil capsules or powdered garlic supplements. Odourless, aged Kyolic garlic is a good choice.

▪ Don't add salt to your food

Don't add salt while you're cooking or to the food on your plate. The only salt we consider healthy is Solo salt, which has half the sodium and lots of potassium and magnesium. Use this in moderation instead.

▪ Cut back on tea and coffee

Don't drink more than one cup of caffeinated or non-caffeinated coffee, or two cups of tea, a day. Instead choose from the wide variety of herbal teas and grain coffees available.

▪ Limit your alcohol

Assuming you don't have a drinking problem, limit your alcohol intake to no more than half a pint of beer or one glass of red wine a day. Ideally, limit your intake to two pints of beer or four glasses of wine a week. If you are unable to limit it to this amount, your H score and your health could suffer.

▪ Reduce your stress

If you are under a lot of stress, or find yourself reacting stressfully much of the time, make a decision to reduce your stress load, both by changing the circumstances that are giving you stress and by changing your attitude. Simple additions to your life, such as yoga, meditation and/or exercise, or seeing a counsellor if you've got some issues to resolve, can make all the difference.

▪ Stop smoking

If you smoke, make a decision to stop, and seek help to do it. There is simply no safe level of smoking as far as homocysteine and your health is concerned. Smoking cigarettes, pipes or cigars is nothing less than slow suicide. The sooner you stop the longer you'll live.

■ Supplement a high-strength multivitamin every day

Take a high-strength multivitamin and mineral supplement, providing at least 25mg of the main B vitamins, 400mcg of folate, 500mcg of B12 (preferably as methylcobalamin) and 25mg of B6, plus A, D, a natural, mixed E and the minerals magnesium, selenium, chromium and zinc. Also supplement 500mg of vitamin C for general health. You'll need TMG too, depending on your H score.

Here are the guidelines.

Nutrient	No risk Below 6	Low risk 6–9	High risk 9–15	Very high risk Above 15
Folate	200mcg	400mcg	1,200mcg	2,000mcg
B12	10mcg	500mcg	1,000mcg	1,500mcg
B6	25mg	50mg	75mg	100mg
B2	10mg	15mg	20mg	50mg
Zinc	5mg	10mg	15mg	20mg
TMG	500mg	750mg	1.5–3g	3–6g

■ Correct oestrogen deficiency

If you are post-menopausal, or have menopausal symptoms or other menstrual irregularities, seek a specialist and check your oestrogen, progesterone and testosterone levels with a hormone saliva test. If you are oestrogen or progesterone deficient, you can correct this with 'natural progesterone' HRT, in the form of a transdermal skin cream. Natural progesterone has none of the associated risks of HRT and your body can make its own oestrogen from progesterone.

Appendix 1

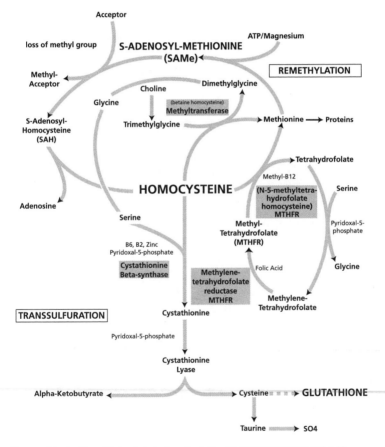

Source: Adapted from Verhoef et al., *American Journal of Epedemiology*, vol. 143 (1996), pp. 845–59

Figure 9. The biochemistry of homocysteine

Appendix 2

Over 100 medical conditions associated with high homocysteine

Note that conditions marked 'high risk' need routine homo-cysteine testing.

Abortions, spontaneous, recurrent and/or early – *high risk*

Accelerated ageing (as a consequence of B vitamin, remethylation, glutathione and SAMe deficiencies)

Alcoholism – *high risk*

Alzheimer's disease – *high risk*

Anaemias (macrocytic/B12 or folate deficiency-induced)

Angina pectoris, coronary atherosclerosis – *high risk*

Ankylosing spondylitis

Apoptosis of brain cells (cell death resulting from the neurotoxic effect of homocysteine, probably due to lack of DNA repair)

Arteriosclerosis (stiffening of walls of arteries)

Arthritis, reactive

Atherosclerosis

Atrial fibrillation

Atrophic gastritis, *Helicobacter pylori*-induced

Autoimmune disease (such as IDDM, ankylosing spondylitis, reactive arthritis and hypothyroidism/Hashimoto's thyroiditis)

Behcet's disease (ulcers in mouth and on genitalia, eye inflammation) with history of thrombosis

Birth defects (such as cleft palate, club foot, neural tube defects, premature birth, urinary tract abnormalities, heart defects, pyloric stenosis) – *high risk*

Bone marrow myelodysplasia/myelodysplastic syndromes (form of cancer similar to leukaemias)

Brain atrophy in normal, healthy elderly patients (increase in brain ventricle to brain tissue ratio)

Breast cancer

Cancers of the colon, thyroid, skin of head and neck, and leukaemias, for example. Also associated with an increase in levels of tumour markers used to measure growth of tumour size

Cervical dysplasia – *high risk*

Chlamydia pneumoniae infection with IgG seropositivity

Cholesterol, elevated (high total & LDL as found in hypothyroidism)

Cirrhosis, fibrosis of the liver

Cleft palate

Club foot

Cobalt deficiency

Coeliac disease – *high risk*

Cognition, lowering in older age (7 to 8 per cent decline)

Colon cancer associated with ulcerative colitis

Copper deficiency

Coronary artery atherosclerosis

Coronary vasospasm

Crohn's disease – *high risk*

Deep vein thrombosis – *high risk*

Dementia, poor cognition

Depression, severe – *high risk*

Diabetes, non insulin dependent (diabetic neuropathy) – *high risk*

Down's syndrome (mothers of babies with the condition)

Epileptics on anti-seizure medications

Epileptic seizures in infants and young children (convulsive status epilepticus)

Fibromyalgia associated with chronic fatigue

Folate deficiency – *high risk*

Gestosis (toxaemia of pregnancy)

Glossitis (inflammation of the tongue)

Glutathione lowering in liver and brain

Heart abnormalities, congenital

Heart attacks/myocardial infarction/atherosclerosis/coronary vasospasm/angina – *high risk*

Heart transplant patients, heart transplant patients with surgical complications, heart patients after corrective surgeries – *high risk*

Helicobacter pylori associated with atherosclerotic strokes – *high risk*

Helicobacter pylori infection associated with stomach and duodenal ulcers

HIV/AIDS – *high risk*

Homocystinuria, a very rare genetic disorder associated with high levels of homocysteine in urine and blood, as well as mental retardation, premature cardiovascular and skeletal diseases and dislocation of lenses

Hypothyroidism

IGF-1 (insulin-like growth factor 1), low levels

Kidney failure, chronic/kidney dialysis patients – *high risk*

Kidney transplant patients

Leukaemia (acute lymphatic leukaemia)

Liver depletion of reduced glutathione, SAMe and/or vitamin E

Lipid peroxidation, lipofuscin (age spots) and low vitamin E levels in liver

Liver disease (alcoholic liver disease, cirrhosis, hepatitis)

Liver transplant patients

Lupus erythematosus with cardiovascular disease

Magnesium deficiency, abnormal metabolism associated with vasospasm

Mental retardation

Methylation/remethylation/transmethylation deficiencies

Methylenetetrahydrofolate genetic mutations (MTHFR) – *high risk*

Methylmalonic acid elevation associated with B12 deficiency

Methylmalonic aciduria with homocystinuria in newborn infant

Migraines with homozygous mutation in the MTHFR gene

Neural tube defect (such as spina bifida) – *high risk*

Neuropsychiatric disorders with B12 deficiency (paresthesias, sensory loss, ataxia, dementia and psychiatric symptoms)

Nitric oxide production inhibition in artery walls

Oestrogen deficiency, post-menopausal state

Optic ischemic neuropathy (in non-arteritic, non-diabetic, younger patients)

Oral dysplasia in elderly who are not tube-fed

Oral lesions in women (glossitis, angular stomatitis)

Orofacial cleft birth defect

Osteoporosis, senile and post-menopausal

Paresthesias (sensations of pricking, tingling, or creeping on the skin)

Parkinson's disease – *high risk*

Penile erection dysfunction

Pernicious anaemia/B12 deficiency (Biermer's disease)

Placental calcification

Polycystic ovary disease

Pre-eclampsia

Pregnancy problems (such as spontaneously recurring abortions, pre-eclampsia, gestosis, placental calcification, premature birth, neural tube defects)

Premature births

Premature death from all common causes (cardiovascular and non-cardiovascular)

Presbyopia

Psoriasis

Psychiatric symptoms (such as depression, dementia, schizophrenia, confusion, sleep disorders)

Pulmonary embolism – *high risk*

Pyloric stenosis in newborns

Raynaud's disease

Restenosis in coronary and peripheral arterial disease following angioplasty – *high risk*

Retinal vein occlusion/thrombosis – *high risk*

Rheumatoid arthritis with history of thrombosis

SAMe (S-adenosylmethionine) deficiency in liver and brain

Schizophrenia with low folate levels, chronic schizophrenia in young males

Sleep apnoea accompanied by cardiovascular disease – *high risk*

Sperm motility reduction

Spina bifida – *high risk*

Squamous cell carcinoma of head and neck

Stomatitis (any of numerous inflammatory diseases of mucosal lining of mouth)

Strokes/cerebral vasospasm/atherosclerosis, including ischemic strokes in children – *high risk*

Sudden death in CAD patients even without acute thrombus

Telomere shortening (associated with accelerated cell ageing, cancer and coronary artery disease)

Thyroid cancer

Thyroiditis, hypothyroidism

Thrombotic disease/abnormal blood clotting – *high risk*

Transient ischemic attacks (TIA) – *high risk*

Ulcerative colitis – *high risk*

Vasospasm, cerebral, associated with abnormal magnesium metabolism

Vasospasm, coronary – *high risk*

Vitamin B12 (cobalamin) stage III deficiency – *high risk*

Vitamin B12 C/D deficiency, an inborn error of B12 metabolism, associated with elevated Hcy, homocysteine, methylmalonic acid, seizures and abnormal EEGs

Vitamin B6 (pyridoxine) deficiency – *high risk*

Vitamin B2 (riboflavin) deficiency – *high risk*

Others . . .

References

Part 1

1. *Newsweek*, news report, 11 August 1997
2. J. Selhub et al., 'Association between plasma homocysteine concentrations and extracranial carotid artery stenosis', *New England Journal of Medicine*, vol. 332, no. 5 (1995), pp. 286–91
3. I. Graham et al., 'Plasma homocysteine as a risk factor for vascular disease', *Journal of the American Medical Association*, vol. 277, no. 22 (1997), pp. 1775–81
4. *Moody's Life Expectation Tables*, US Department of Health and Human Resources (1996)
5. S. E. Vollset et al., 'Plasma total homocysteine and cardiovascular and non-cardiovascular mortality: the Hordal and Homocsyteine Study', *American Journal of Clinical Nutrition*, vol. 74, no. 1 (2001), pp. 130–6
6. P. Lichtenstein et al., 'Environmental and heritable factors in the causation of cancer – analyses of cohorts of twins from Sweden, Denmark and Finland', *New England Journal of Medicine*, vol. 343, no. 2 (2002), pp. 78–85
7. M. Toshifumi et al., 'Elevated plasma homocysteine levels and risk of silent brain infarction in elderly people', *Stroke*, vol. 32 (2001), pp. 1116
8. S. Seshadri et al., 'Plasma homocysteine as a risk factor for dementia and Alzheimer's disease', *New England Journal of Medicine*, vol. 346, no. 7 (2002), pp. 476–83
9. K. Dalery et al., 'Homocysteine and coronary artery disease in French Canadian subjects: relation with vitamins B12, B6, pyridoxal phosphate, and folate', *American Journal of Cardiology*, vol. 75, (1995) pp. 1107–11
10. A. Cortelezzi et al., 'Hyperhomocysteinemia in myelodysplastic syndromes: specific association with autoimmunity and cardiovascular disease', *Leukemia & Lymphoma*, vol. 41 (2001), pp. 147–50

11. M. D. Silverman et al., 'Homocysteine upregulates vascular cell adhesion molecule-1 expression in cultured human aortic endothelial cells and enhances monocyte adhesion', *Thrombosis & Vascular Biology*, vol. 22 (2002), pp. 587–92

12. M. F. McCarty, 'P-regulation of endothelial nitric oxide activity as a central strategy for prevention of ischemic stroke – just say NO to stroke!' *Medical Hypotheses*, vol. 55 (2000), pp. 386–403

13. B. Mutus et al., 'Homocysteine-induced inhibition of nitric oxide production in platelets: a study on healthy and diabetic subjects', *Diabetologia*, vol. 44, no. 8 (2001), pp. 979–82

14. N. Li et al., 'Effects of homocysteine on intracellular nitric oxide and superoxide levels in the renal arterial endothelium', *American Journal of Physiology – Heart Circulation Physiology*, vol. 283 (2002), pp. 1237–43

15. D. Benton and G. Roberts, 'Effect of vitamin and mineral supplementation on intelligence of a sample of school children', *Lancet*, vol. 1 (1988), pp. 140–3

16. M. G. Signorello, R. Pascale and G. Leoncini, 'Effect of homocysteine on arachidonic acid release in human platelets', *European Journal of Clinical Investigation*, vol. 32 (2002), pp. 279–84

17. Dr I. I. Kruman et al., 'Homocysteine and folate deficiency impair DNA repair in brain neurons and sensitize cells to toxicity by amyloid bodies', *Journal of Neuroscience*, vol. 22 (2002), pp. 1752–62

18. G. Kempermann and F. Gage, 'New nerve cells for the adult brain', *Scientific American*, vol. 12, no. 1 (2002), pp. 38–44

19. K. R. Dimitrova et al., 'Estrogen and homocysteine', *Cardiovascular Research*, vol. 53, no. 3 (2002), pp 577–88

20. Hong Tao et al., 'Effects of low-dose conjugated estrogen on plasma homocysteine level in patients with coronary heart disease', Department of Cardiology, Peking University First Hospital, sponsored by the fund of Ministry of Public Health of China, no. 98-1-251

21. C. W. Fetrow and J. R. Avila, 'Efficacy of the dietary supplement S-adenosyl-L-methionine', *Annals of Pharmacotherapy*, vol. 35 (2001), pp. 1414–25

22. K. L. Soeken et al., 'Safety and efficacy of S-adenosylmethionine (SAMe) for osteoarthritis', *Journal of Family Practice*, vol. 51 (2002), pp. 425–30

23. S. Jacobsen et al., 'Oral S-adenosylmethionine in primary fibromyalgia: Double-blind clinical evaluation', *Scandinavian Journal of Rheumatology*, vol. 20 (1991), pp. 294–302 and A. Tavoni et al., 'Evaluation of S-adenosylmethionine in primary fibromyalgia: A double-blind crossover study', *American Journal of Medicine*, vol. 83, Suppl. 5A (1978), pp. 107–10

24. S. M. Tsai et al., 'Effects of S-adenosyl-L-methionine on liver damage in experimental obstructive jaundice', *Journal of Medical Science*, vol. 17 (2001), pp. 455–60

25. J. P. De La Cruz et al., 'Effects of S-adenosyl-L-methionine on lipid peroxidation and glutathione levels in rat brain slices exposed to reoxygenation after oxygen-glucose deprivation', *Neuroscience Letter*, vol. 318 (2002) pp. 103–7

Part 2

1. J. Selhub et al., 'Association between plasma homocysteine concentrations and extracranial carotid artery stenosis', *New England Journal of Medicine*, vol. 332, no. 5 (1995), pp. 286–91

2. D. S. Wald and J. K. Morris, 'Homocysteine and cardiovascular disease: evidence on causality from a meta-analysis', *British Medical Journal*, vol. 325 (2002), pp. 1202

3. J. Selhub et al., 'Association between plasma homocysteine concentrations and extracranial carotid artery stenosis', *New England Journal of Medicine*, vol. 332, no. 5 (1995), pp. 286–91

4. Nygard et al., 'Plasma homocysteine levels and mortality in patients with coronary artery disease', *New England Journal of Medicine,* vol. 337, no. 4 (1997), pp. 230–6

5. K. Robinson et al., 'Hyperhomocysteinemia and Low Pyridoxal Phosphate: Common and Independent Reversible Risk Factors for Coronary Artery Disease', *Journal of Circulation* , vol. 92 (1995), pp. 2825–30

6. P. Verhoef et al., 'Homocysteine metabolism and risk of myocardial infarction: relation with vitamins B6, B12, and folate', *American Journal of Epidemiology,* vol. 143, no. 9 (1996), pp. 845–59

7. F. Andreotti et al., 'Homocysteine and arterial occlusive disease: a concise review', *Cardiologia*, vol. 44, no. 4 (1999), pp. 341–5

8. D. Bunout et al., 'Serum homocysteine levels in healthy Chilean adults', *Revista Medica de Chile*, vol. 126 (1998), pp. 905–10

9. J. J. Janson et al., 'Prevalence of hyperhomocysteinemia in an elderly population', *American Journal of Hypertension*, vol. 15 (2002), pp. 394–7

10. F. Andreotti et al., 'Homocysteine and arterial occlusive disease: a concise review', *Cardiologia*, vol. 44, no. 4 (1999), pp. 341–5

11. K. Dalery et al., 'Homocysteine and coronary artery disease in French Canadian subjects: relation with vitamins B12, B6, pyridoxal phosphate, and folate', *American Journal of Cardiology*, vol. 75. no. 16 (1995) pp. 1107–11

12. R. Carmel et al., 'Hyperhomocysteinemia and cobalamin deficiency in young Asian Indians in the United States', *American Journal of Hematology*, vol. 70 (2002), pp. 107–14

13. E. Okada et al., 'Hyperhomocysteinemia is a risk factor for coronary arteriosclerosis in Japanese patients with type 2 diabetes', *Diabetes Care*, vol. 22, no. 3 (1999), pp. 484–90

14. M. Rauh et al., 'Homocysteine concentrations in a German cohort of 500 individuals: reference ranges and determinants of plasma levels in healthy children and their parents', *Amino Acids*, vol. 20, no. 4 (2001), pp. 409–18

15. R. M. Ortega et al., 'Homocysteine levels in elderly Spanish people: influence of pyridoxine, vitamin B12 and folic acid intakes', *Journal of Nutrition, Health and Aging*, vol. 6 (2002), pp. 69–71

16. J. D. Kark et al., 'Plasma homocysteine and parental myocardial infarction in young adults in Jerusalem', *Circulation*, vol. 105 (2002), pp. 2725–9

17. L. L. Kwan, O. I. Bermudez and K. L. Tucker, 'Low vitamin B-12 intake and status are more prevalent in Hispanic older adults of Caribbean origin than in neighborhood-matched non-Hispanic whites', *Journal of Nutrition*, vol. 132 (2002), pp. 2059–64

18. J. Wang et al., 'The distribution of serum homocysteine and its associated factors in a population of 1,168 subjects in Beijing area', *Zhonghua Liu Xing Bing Xue Za Zhi*, vol. 23 (2002), pp. 32–5

19. K. G. Rowley et al., 'Improvements in circulating cholesterol, antioxidants, and homocysteine after dietary intervention in an Australian Aboriginal community', *American Journal of Clinical Nutrition*, vol. 74, no. 4 (2001), pp. 442–8

20. M. Klerk et al., 'MTHFR 677C–T Polymorphism and Risk of Coronary Heart Disease: A Meta-analysis', *Journal of the American Medical Association*, vol. 288, no. 16 (2002), pp. 2023–31

21. N. Rosenberg et al., 'The frequent 5,10-methylenetetrahydrofolate reductase C677T polymorphism is associated with a common haplotype in whites, Japanese, and Africans', *American Journal of Human Genetics*, vol. 70, no. 3 (2002), pp. 758–62

22. O. Stanger et al., 'Prevalence of hyperhomocyseinemia in patients requiring coronary artery bypass', *Internet Journal of Thoracic and Cardiovascular Surgery*, vol. 2, no. 1 (1997)

23. L. Rebocho, C. Barata and I. Henriques, 'Hyperhomocysteinemia and ischemic stroke: easy to diagnose, easy to treat', presented at the 2002 European Stroke Conference (Risk Factors and Etiology) on 30 May 2002 in Geneva, Switzerland

24. R. van der Griend, D. H. Biesma and J. D. Banga, 'Postmethionine-load homocysteine determination for the diagnosis hyperhomocysteinaemia and efficacy of homocysteine lowering treatment regimens', *Vascular Medicine*, vol. 7 (2002), pp. 29–33

Part 3

1. S. E. Vollset et al., 'Plasma total homocysteine and cardiovascular and non-cardiovascular mortality: the Hordaland Homocysteine Study', *American Journal of Clinical Nutrition*, vol. 74, no. 1 (2001), pp. 130–6

2. D. Xu, R. Neville and T. Finkel, 'Homocysteine accelerates endothelial cell senescence', *FEBS Letter*, vol. 470 (2000), pp. 20–4

3. N. J. Samani et al., 'Telomere shortening in atherosclerosis', *Lancet*, vol. 16, no. 359/9310 (2002), pp. 976–7

4. A. M. Fenech, *New York Academy of Science Journal*, vol. 854 (1998), pp. 23–36

5. D. S. Wald and J. K. Morris, 'Homocysteine and cardiovascular disease: evidence on causality from a meta-analysis', *British Medical Journal*, vol. 325 (2002), pp. 1202

6. M. Clerk et al., 'MTHFR 677C>T polymorphism and risk of coronary heart disease', *Journal of the American Medical Association*, vol. 288, no. 16 (2002) pp. 2023–31

7. A. Gupta et al., 'High homocysteine, low folate, and low vitamin B6 concentrations', *Transplantation*, vol. 65, no. 4 (1998), pp. 544–50

8. G. Schnyder et al., 'Association of plasma homocysteine with restenosis after percutaneous coronary angioplasty', *European Heart Journal*, vol. 23 (2002), pp. 726–33

9. G. Schnyder et al. 'Effect of homocysteine-lowering therapy with folic acid, B12 and B6 on clinical outcome after percutameous intervention', *Journal of the American Medical Association*, vol. 22, no. 8 (2002) pp. 973–9

10. B. T. Altura and B. M. Altura, 'Magnesium in cardiovascular biology', *Scientific American*, May/June 1995, pp. 28–36

11. W. J. Mroczek, W. R Lee and M. E. Davidov, 'Effect of magnesium sulfate on cardiovascular hemodynamics', *Angiology*, vol. 28, no. 10 (1997), pp. 720–4

12. B. T. Altura and B. M. Altura, 'Magnesium in cardiovascular biology', *Scientific American*, May/June 1995, pp. 28–36

13. W. Li et al., 'Homocysteine results in magnesium depletion from vascular smooth muscles', *Neuroscience Letter*, vol. 274 (1999), pp. 83–6

14. H. Morita et al., 'Diet-induced mild hyperhomocysteinemia and increased salt intake diminish vascular endothelial function in a synergistic manner', *Journal of Hypertension*, vol. 20 (2002), pp. 55–62

15. N. Li et al., 'Effects of homocysteine on intracellular nitric oxide and superoxide levels in the renal arterial endothelium', *American Journal of Physiology – Heart and Circulation Physiology*, vol. 283 (2002), pp. 1237–43

16. W. Y. Fu et al., 'Homocysteine attenuates hemodynamic responses to nitric oxide in vivo', *Atherosclerosis*, vol. 161 (2002), pp. 169–76

17. J. D. Symons et al., 'Hyperhomocysteinemia evoked by folate depletion: effects on coronary and carotid arterial function', *Arteriosclerosis, Thrombosis and Vascular Biology*, vol. 22, no. 5 (2002), pp. 772–80

18. A. G. Bostom et al., 'Nonfasting plasma total homocysteine levels and stroke incidence in elderly persons: The Framingham Study', *Annals of Internal Medicine*, vol. 131 (1999 131), pp. 352–5

19. P. K. Sarkar and L. A. Lambert, 'Aetiology and treatment of hyperhomocysteinaemia causing ischaemic stroke', *International Journal of Clinical Practice*, vol. 55, no. 4 (2001), pp. 262–8

20. H. Shimizu et al., 'Plasma homocyst(e)ine concentrations and the risk of subtypes of cerebral infarction: The Hisayama Study', *Cerebrovascular Disease*, vol. 13, no. 1 (2002), pp. 9–15.

21. T. Matsui et al., 'Elevated plasma homocysteine levels and risk of silent brain infarction in elderly people', *Stroke*, vol. 32, no. 5 (2001), pp. 1116–9

22. A. T. Hirsch et al., 'Peripheral Arterial Disease Detection, Awareness, and Treatment in Primary Care', *Journal of the American Medical Association*, vol. 286 (2001), pp. 1317–24

23. S. de Jong, 'Hyperhomocysteinaemia in patients with peripheral arterial occlusive disease', *Clinical Chemistry and Laboratory Medicine*, vol. 39 (2001), pp. 714–16

24. B. A. Brown et al., 'Homocysteine: a risk factor for retinal venous occlusive disease', *Ophthalmology*, vol. 109 (2002), pp. 287–90

25. K. Aksu et al., 'Hyperhomocysteinaemia in Behcet's disease', *Rheumatology*, vol 40, no. 6 (2001), pp. 687–90

26. B. Kuch et al., 'Associations between homocysteine and coagulation factors – a cross-sectional study in two populations of central Europe', *Thrombosis Research*, vol. 103, no. 4 (2001), pp. 265–73

27. L. L. Wu and J. T. Wu, 'Hyperhomocysteinemia is a risk factor for cancer and a new potential tumor marker', *Clinical Chimica Acta*, vol. 322 (2002), pp. 21–8

28. A. Cortelezzi et al., 'Hyperhomocysteinemia in myelodysplastic syndromes: specific association with autoimmunity and cardiovascular disease', *Leukemia & Lymphoma*, vol. 41, nos. 1–2 (2001), pp. 147–50

29. S. W. Thomson et al., 'Correlates of total plasma homocysteine: folic acid, copper, and cervical dysplasia', *Nutrition*, vol. 6 (2000), pp. 411–16

30. R. Prinz-Langenohl, I. Fohr and K. Pietrzik, 'Beneficial role for folate in the prevention of colorectal and breast cancer', *European Journal of Nutrition*, vol. 40 (2001), pp. 98–105

31. B. Shannon, S. Gnanasampanthan, J. Beilby and B. Iacopetta, 'A polymorphism in the methylenetetrahydrofolate reductase gene predisposes to colorectal cancers with microsatellite instability', Department of Surgery, University of Western Australia, awaiting publication

32. J. E. Goodman et al., 'COMT genotype, micronutrients in the folate metabolic pathway and breast cancer risk', *Carcinogenesis*, vol. 22 (2001), pp. 1661–5

33. M. Buysschaert et al., 'Micro- and macrovascular complications and hyperhomocysteinaemia in type 1 diabetic patients', *Diabetes & Metabolism*, vol. 6 (2001), pp. 655–9

34. J. M. Kaye et al., *Clinical Science*, vol. 102 (2002), pp. 631–7

35. S Gallistl et al., 'Insulin is an independent correlate of plasma homocysteine levels in obese children and adolescents', *Diabetes Care*, vol. 23 (2000), pp. 1348–52

36. D. A. de Luis et al., *Diabetes Research and Clinical Practice*, vol. 55 (2002), pp. 185–90

37. B. Mutus et al., 'Homocysteine-induced inhibition of nitric oxide production in platelets: a study on healthy and diabetic subjects', *Diabetologia*, vol. 44, no. 8 (2001), pp. 979–82

38. *Reuters World Report*, 27 April 1998

39. J. W. Green et al., 'Homocysteine, vitamin B6, and vascular disease in AD patients', *Neurology*, vol. 58 (2002), pp. 1471–5

40. S. Seshadri et al., 'Plasma homocysteine as a risk factor for dementia and Alzheimer's disease', *New England Journal of Medicine*, vol. 346, no. 7 (2002), pp. 466–8

41. P. S. Sachdev et al., *Neurology*, vol. 58 (2002), pp. 1539–41

42. T. Bottiglieri et al., 'Plasma total homocysteine levels and the C677T mutation in the methylenetetrahydrofolate reductase (MTHFR) gene: a study in an Italian population with dementia', *Mechanisms of Ageing and Development*, vol. 122, no. 16 (2001), pp. 2013–23

43. K. Eto et al., 'Brain hydrogen sulfide is severely decreased in Alzheimer's disease', *Biochemical and Biophysical Research Communications*, vol. 293 (2002), pp. 1485–8

44. S. J. Duthie et al., 'Homocysteine, B vitamin status, and cognitive function in the elderly', *American Journal of Clinical Nutrition*, vol. 75, no. 5 (2002), pp. 908–13

45. J. G. Hollowell et al., 'Serum Thyrotropin, Thyroxine, and Thyroid Antibodies in the United States Population (1988 to 1994): National Health and Nutrition Examination Survey (NHANES III)', *Journal of Clinical Endocrinology & Metabolism*, vol. 87, no. 2 (2002), pp. 486–8

46. W. I. Hussein et al., 'Normalization of hyperhomocysteinemia with L-thyroxine in hypothyroidism', *Annals of Internal Medicine*, vol. 131, no. 5 (1999), pp. 348–51

47. M. Bicikova et al., 'Effect of treatment of hypothyroidism on the plasma concentrations of neuroactive steroids and homocysteine', *Clinical Chemistry and Laboratory Medicine*, vol. 39, no. 8 (2001), pp. 753–7

48. J. C. Toft and H. Toft, 'Hyperhomocysteinemia and hypothyroidism', *Ugeskr Laeger* (Denmark), vol. 163, no. 34 (2001), pp. 4593–4

49. M. S. Morris et al., 'Hyperhomocysteinemia and hypercholesterolemia associated with hypothyroidism in the third US National Health and Nutrition Examination Survey', *Atherosclerosis*, vol. 155, no. 1 (2001), pp. 195–200

50. B. Catargi et al., 'Homocysteine, hypothyroidism, and the effect of thyroid hormone replacement', *Thyroid*, vol. 9 (1999), pp. 1163–6

51. *The World Health Report 2001 – Mental Health: New Understanding, New Hope*, World Health Organization (2001)

52. M. Fava et al., 'Folate, vitamin B12 and homocysteine in major depressive disorder', *American Journal of Pyschiatry*, vol. 154 (1997), pp. 426–8

53. P. S. Godfrey et al., 'Enhancement of recovery from psychiatric illness by methylfolate', *Lancet*, vol. 336, no. 8712 (1990), pp. 392–5

54. R. Crellin et al., 'Folates and psychiatric disorders: Clinical potential', *Drugs*, vol. 45, no. 5 (1993), pp. 623–36

55. M. Murray, *Encyclopaedia of Natural Supplements* (1996), Prima Publications

56. T. Bottiglieri, M. Laundy et al., 'Homocysteine, folate, methylation, and monoamine metabolism in depression', *Journal of Neurology, Neurosurgery & Psychiatry*, vol. 69 (2000), pp. 228–32

57. B. Regland et al., 'Homocysteinemia is a common feature of schizophrenia', *Journal of Neural Transmission, General Section*, vol. 100, no. 2 (1995), pp. 165–9

58. J. Levine et al., 'Elevated homocysteine levels in young male patients with [chronic] schizophrenia', *American Journal of Psychiatry*, vol. 159, no. 10 (2002), pp. 1790–2

59. R. Regland et al., 'Homocysteinemia and schizophrenia as a case of methylation deficiency', *Neural Transmission, General Section*, vol. 98, no. 2 (1994), pp. 143–52

60. E. Susser et al., 'Schizophrenia and impaired homocysteine metabolism: a possible association', *Biological Psychiatry*, vol. 44, no. 2 (1998), pp. 141–3

61. J. Lindenbaum et al., 'Neuropsychiatric disorders caused by cobalamin deficiency in the absence of anemia or macrocytosis', *English Journal of Medicine*, vol. 318 (1988), pp. 1720–8

62. R. B. D'Agostino et al., 'Plasma homocysteine as a risk factor for dementia and Alzheimer's disease', *New England Journal of Medicine*, vol. 346, no. 7 (2002), pp. 476–83

63. W. Duan, M. P. Mattson et al., 'Dietary folate deficiency and elevated homocysteine levels endanger dopaminergic neurons in models of Parkinson's disease', *Journal of Neurochemistry*, vol. 80, no. 1 (2002), pp. 101–10

64. S. E. Vollset et al., 'Plasma total homocysteine, pregnancy complications, and adverse pregnancy outcomes: the Hordaland Homocysteine study', *American Journal of Clinical Nutrition*, vol. 71, no. 4 (2000), pp. 962–8

65. P. de Marco, 'Polymorphisms in genes involved in folate metabolism as risk factors for NTDs', *European Journal of Pediatric Surgery*, supp. 1 (2001), pp. S14–17

66. W. L. Nelen et al, 'Homocysteine and folate levels as risk factors for recurrent early pregnancy loss', *Obstetrics & Gynecology*, vol. 95, no. 4 (2000), pp. 519–24

67. R. Carmi et al., 'Spontaneous abortion – high risk factor for neural tube

defects in subsequent pregnancy', *American Journal of Medical Genetics*, vol. 51 (1994), pp. 93–7

68. H. Zetterberg et al., 'Increased frequency of combined methylenetetrahydrofolate reductase C677T and A1298C mutated alleles in spontaneously aborted embryos', *European Journal of Human Genetics*, vol. 10, no. 2 (2002), pp. 113–18

69. W. L. Nelen et al., 'Homocysteine and folate levels as risk factors for recurrent early pregnancy loss', *Obstetrics & Gynecology*, vol. 95, no. 4 (2000), pp. 519–24

70. T. O. Scholla and W. G. Johnson, 'Folic acid: influence on the outcome of pregnancy', *American Journal of Clinical Nutrition*, vol. 71 (2000), pp. 129–303

71. M. Reznikoff-Etiévant et al., 'Low Vitamin B(12) level as a risk factor for very early recurrent abortion', *European Journal of Obstetrics, Gynecology, and Reproductive Biology*, vol. 104 (2002), p. 156

72. M. T. Steen et al., 'Neural-tube defects are associated with low concentrations of cobalamin (vitamin B12) in amniotic fluid', *Prenatal Diagnosis*, vol. 18 (1998), pp. 545–55

73. A. G. Ronnenberg et al., 'Preconception folate and vitamin B(6) status and clinical spontaneous abortion in Chinese women', *Obstetric Gynecology*, vol. 100 (2002), pp. 107–13

74. H. Böhles et al., 'Maternal plasma homocysteine, placenta status and docosahexaenoic acid concentration in erythrocyte phospholipids of the newborn', *European Journal of Pediatrics*, vol. 158 (1999), pp. 243–6

75. A. M. Molloy et al., 'Maternal and fetal plasma homocysteine concentrations at birth: the influence of folate, vitamin B12, and the 5,10-methylenetetrahydrofolate reductase 677C—>T variant', *American Journal of Obstetrics and Gynecology*, vol. 186 (2002), pp. 499–503

76. L. I. Al-Gazali et al., 'Abnormal folate metabolism and genetic polymorphism of the folate pathway in a child with Down syndrome and neural tube defect', *American Journal of Medical Genetics*, vol. 103, no. 2 (2001), pp. 128–32

77. M. Pogribna et al., 'Homocysteine metabolism in children with Down syndrome: in vitro modulation', *American Journal of Human Genetics*, vol. 69, no. 1 (2001), pp. 88–95

78. R. Biancheri et al., 'Early-onset cobalamin C/D deficiency: epilepsy and electroencephalographic features', *Epilepsia*, vol. 43 (2002), pp. 616–22

79. B. V. Sastry and V. E. Janson, 'Depression of human sperm motility by inhibition of enzymatic methylation', *Biochemical Pharmacology*, vol. 32 (1983), pp. 1423–32

80. G. Loverro, 'The plasma homocysteine levels are increased in polycystic ovary syndrome', *Gynecologic and Obstetric Investigation*, vol. 53 (2002), pp. 157–62

81. A. Hernanz et al., 'Increased plasma levels of homocysteine and other thiol compounds in rheumatoid arthritis women', *Clinical Biochemistry*, vol. 32, no. 1 (1999), pp. 65–70

82. B. Seriolo, D. Fasciolo, A. Sulli and M. Cutolo, 'Homocysteine and antiphospholipid antibodies in rheumatoid arthritis patients: relationships with thrombotic events', *Clinical and Experimental Rheumatology,* vol. 19, no. 5 (2001), pp. 561–4

83. X. M. Gao et al., 'Homoycysteine, Ankylosing Spondylitis and Reactive Arthritis: Homocysteine modification of HLA antigens and its immunological consequences', *European Journal of Immunology,* vol. 26 (1996), pp. 1443–50

84. M. A. Flynn et al., 'The effect of folate and cobalamin on osteoarthritic hands', *Journal of the American College of Nutrition,* vol. 13, no. 4 (1994), pp. 351–6

85. J. M. Ellis, *Free of Pain: A Proven and Inexpensive Treatment for Specific Types of Rheumatism,* Southwest Publishing (1983)

86. M. Miyao et al., 'Association of methylenetetrahydrofolate reductase (MTHFR) polymorphism with bone mineral density in postmenopausal Japanese women', *Calcified Tissue International,* vol. 66 (2000), pp. 190–4

87 K. R. Dimitrova et al., 'Estrogen and homocysteine,' *Cardiovascular Resident,* vol. 53 (2002), pp. 577–88

88. V. Mijatovic and M. J. van der Mooren, 'Homocysteine in postmenopausal women and the importance of hormone replacement therapy', *Clinical Chemistry and Laboratory Medicine,* vol. 39 (2001), pp. 764–767

89. B. Regland et al., 'Increased concentrations of homocysteine in the cerebrospinal fluid in patients with fibromyalgia and chronic fatigue syndrome', *Scandinavian Journal of Rheumatology,* vol. 26 (1997), pp. 301–7

90. H. Kowa et al., 'The homozygous C677T mutation in the methylenetetrahydrofolate reductase gene is a genetic risk factor for migraine', *American Journal of Medical Genetics,* vol. 96, no. 6 (2000), pp. 762–4

91. P. Coll et al., *Nor Laegeforen* (Norway), vol. 119 (1999), pp. 3577–9

92. A. Tamura, T. Fujioka and M. Nasu, 'Relation of *Helicobacter pylori* infection to plasma vitamin B12, folic acid, and homocysteine levels in patients who underwent diagnostic coronary arteriography', *American Journal of Gastroenterology,* vol. 97 (2002), pp. 861–6

93. Antonio Piertroiusti et al., 'Cytotoxin-associated gene-A-positive *Helicobacter pylori* strains are associated with atherosclerotic stroke', *Circulation,* vol. 106 (2002), pp. 580–4

Part 4

1. G. H. J. Boers, A. G. H. Smals, F. J. M. Trijbels et al., 'Heterozygosity for homocystinuria in premature peripheral and cerebral occlusive arterial disease', *New England Journal of Medicine,* vol. 313 (1985) pp. 709–15

2. R. Z. Stolzenberg-Solomon et al., 'American Society for Clinical Nutrition association of dietary protein intake and coffee consumption with serum homocysteine concentrations in an older population', *American Journal of Clinical Nutrition,* vol. 69, no. 3 (1999), pp. 467–75

3. M. S. Nenseter et al., 'Effect of Norwegian fish powder on risk factors for coronary heart disease among hypercholesterolemic individuals,' *Nutrition Metabolism and Cardiovascular Disease*, vol. 10 (2000), pp. 323–30

4. H. Grundt et al., 'Atherothrombogenic risk modulation by n-3 fatty acids was not associated with changes in homocysteine in subjects with combined hyperlipidaemia', *Thrombosis and Haemostasis*, vol. 81 (1999), pp. 561–5

5. D. J. A. Jenkins et al., 'Effects of high- and low-isoflavone soyfoods on blood lipids, oxidized LDL, homocysteine, and blood pressure in hyperlipidemic men and women', *American Journal of Clinical Nutrition*, vol. 76, no. 2 (2002), pp. 365–372

6. S. Tonstad, K. Smerud and L. Høie, 'A comparison of the effects of 2 doses of soy protein or casein on serum lipids, serum lipoproteins, and plasma total homocysteine in hypercholesterolemic subjects', *American Journal of Clinical Nutrition*, vol. 76, no. 1 (2002), pp. 78–84

7. C. J. Hung et al., 'Buddhist lacto-vegetarians have lower blood B12 levels and higher homocysteine vs omnivores in Taiwan study. Plasma homocysteine levels in Taiwanese vegetarians are higher than those of omnivores', *Journal of Nutrition*, vol. 132, no. 2 (2002), pp. 152–8

8. D. Mezzano et al., 'Vegetarians and cardiovascular risk factors: hemostasis, inflammatory markers and plasma homocysteine', *Thrombosis and Haemostasis*, vol. 81 (1999), pp. 913–17

9. D. Mezzano et al., 'Cardiovascular risk factors in vegetarians: normalization of hyperhomocysteinemia with vitamin B(12) and reduction of platelet aggregation with n-3 fatty acids', *Thrombosis Research*, vol. 100 (2000), pp. 153–60

10. W. M. Broekmans, I. A. Klopping-Ketelaars et al., 'Fruits and vegetables increase plasma carotenoids and vitamins and decrease homocysteine in humans', *Journal of Nutrition*, vol. 130 (2000), pp. 1578–83

11. M. R. Malinow et al., 'Reduction of plasma homocyst(e)ine levels by breakfast cereal fortified with folate in patients with coronary heart disease', *New England Journal of Medicine*, vol. 338 (1998), pp. 1009–15

12. L. J. Riddell et al., 'Dietary strategies for lowering homocysteine concentrations', *American Journal of Clincial Nutrition*, vol. 71 (2000), pp. 1448–54

13. R. Abraham et al., 'Diets of Asian pregnant women in Harrow: iron and vitamins', *Human Nutrition: Applied Nutrition*, vol. 41 (1987), pp. 164–73

14. J. H. Matthews et al., 'Megaloblastic anemia in vegetarian Asians', *Clinical and Laboratory Haematology*, vol. 6 (1984), pp. 1–7

15. Y. Yeh et al., 'Garlic extract reduces plasma concentration of homocysteine in rats rendered folate deficient', *The FASEB Journal: Official Publication of the Federation of American Societies for Experimental Biology*, vol. 13, no. 4 (1999), pp. 209–12

16. H. Morita et al., 'Homocysteine is particularly harmful to arteries in the presence of a high salt intake. Diet-induced mild hyperhomocysteinemia and increased salt intake diminish vascular endothelial function in a synergistic manner', *Journal of Hypertension*, vol. 20 (2002), pp. 55–62

17. J. M. Geleijnse et al., 'Reduction in blood pressure with low sodium, high potassium, high magnesium salt in older subjects with mild to moderate hypertension', *British Medical Journal*, vol. 309, no. 6952 (1994), pp. 436–40

18. R, M. Fleming et al. 'The effect of high-, moderate-, and low-fat diets on weight loss and cardiovascular disease risk factors', *Preventive Cardiology*, vol. 5 (2002), pp. 110–18

19. M. J. Grubben et al., 'Unfiltered coffee increases plasma homocysteine concentrations in healthy volunteers: a randomized trial', *American Journal of Clinical Nutrition*, vol. 71, no. 2 (2000), pp. 480–4

20. P. Verhoef et al., 'Contribution of caffeine to the homocysteine-raising effect of coffee: A randomized controlled trial in humans', *American Journal of Clinical Nutrition*, vol. 76, no. 6 (2002), pp. 1244–8

21. M. R. Olthof et al., 'Consumption of high doses of chlorogenic acid, present in coffee, or of black tea increases plasma total homocysteine concentrations in humans', *Journal of Clinical Nutrition*, vol. 73, no. 3 (2001), pp. 532–8

22. R. Z. Stolzenberg-Solomon et al., 'Association of dietary protein intake and coffee consumption with serum homocysteine concentrations in an older population', *American Journal of Clinical Nutrition*, vol. 69, no. 3 (1999), pp. 467–75

23. P. F. Jacques et al., 'Determinants of plasma total homocysteine concentration in the Framingham Offspring cohort', *American Journal of Clinical Nutrition*, vol. 73, no. 3 (2001), pp. 613–21

24. S. Bleich et al., 'Moderate alcohol consumption in social drinkers raises plasma homocysteine levels: a contradiction to the 'French Paradox'?', *Alcohol*, vol. 36 (2001), pp. 189–92

25. T. Truelsen et al., 'Amount and type of alcohol and risk of dementia: the Copenhagen City Heart Study', *Neurology*, vol. 59, no. 9 (2002), pp. 1313–19

26 M. S. van der Gaag et al., 'Effect of consumption of red wine, spirits, and beer on serum homocysteine', *Lancet*, vol. 355 (2000), p. 1522

27. O.Mayer et al., 'A population study of the influence of beer consumption on folate and homocysteine concentrations', *European Journal of Clinical Nutrition*, vol. 55 (2001), pp. 605–9

28. J. B. Dixon et al., 'Reduced plasma homocysteine in obese red wine consumers: a potential contributor to reduced cardiovascular risk status', *European Journal of Clinical Nutrition*, vol. 56 (2002), pp. 608–14

29. S. Bleich et al., 'Homocysteine and alcoholism', *Journal of Neural Transmission*, Suppl. (2000), pp. 187–96

30. A. J. Barak et al., 'Betaine, ethanol, and the liver: a review', *Alcohol*, vol. 13 (1996), pp. 395–8

31. M. J. de la Vega, F. Santolaria et al., 'High prevalence of hyperhomocysteine-mia in chronic alcoholism: the importance of the thermolabile form of the enzyme methylenetetra-hydrofolate reductase (MTHFR)', *Alcohol*, vol. 25, no. 2 (2001), pp. 59–67

32. A. J. Barak et al., 'Chronic ethanol consumption increases homocysteine accumulation in hepatocytes', *Alcohol*, vol. 25 (2001), pp. 77–81

33. P. H. Black and L. D. Garbutt, 'Stress, inflammation and cardiovascular disease', *Journal of Psychosomatic Research*, vol. 52 (2002), pp. 1–23

34. C. M. Stoney and T. O. Engebretson, 'Plasma homocysteine concentrations are positively associated with hostility and anger', *Life Science*, vol. 66 (2000), pp. 2267–75

35. I. Kato et al, 'Epidemiologic correlates of serum folate and homocysteine levels among users and non-users of vitamin supplement', *International Journal of Vitamin and Nutrition Research*, vol. 69 (1999), pp. 322–9

36. J. H. Stein et al., 'Smoking cessation, but not smoking reduction, reduces plasma homocysteine levels', *Clinical Cardiolology*, vol. 25 (2002), pp. 23–6

37. J. W. van Wersch et al., 'Folate, Vitamin B(12), and homocysteine in smoking and non-smoking pregnant women', *European Journal of Obstetrics, Gynecology, and Reproductive Biology*, vol. 103 (2002), pp. 18–21

38. J. B. Dixon, M. E. Dixon and P. E. O'Brien, 'Homocysteine levels with weight loss after lap-band surgery: higher folate and vitamin B12 levels required to maintain [safer] homocysteine level', *International Journal of Obesity and Related Metabolic Disorders*, vol. 25 (2001), pp. 219–27

39. B. F. Henning et al., 'Vitamin supplementation during weight reduction – favorable effect on homocysteine metabolism', *Research in Experimental Medicine* (Berlin), vol. 198 (1998), pp. 37–42

40. Hong Tao et al., 'Effects of low-dose conjugated estrogen on plasma homo-cysteine level in patients with coronary heart disease', Department of Cardiology, Peking University First Hospital, sponsored by the fund of Ministry of Public Health of China, no. 98-1-251

41. S. Giri, P. D. Thompson, P. Taxel et al., 'Oral estrogen improves serum lipids, homocysteine and fibrinolysis in elderly men', *Atherosclerosis*, vol. 137 (1998), pp. 359–66

42. V. Beral et al., 'Evidence from randomised trials on the long-term effects of hormone replacement therapy', *Lancet*, vol. 360, no. 9337 (2002), pp. 942–4

43. C. Desouz et al., 'Drugs affecting homocysteine metabolism: impact on cardiovascular risk', *Drugs*, vol. 62 (2002), pp. 605–16 and R. Fijnheer et al., 'Homocysteine, methylenetetrahydrofolate reductase polymorphism, antiphospholipid antibodies, and thromboembolic events in systemic lupus

erythematosus: a retrospective cohort study', *Journal of Rheumatology*, vol. 25 (1998), pp. 1737–42

44. C. Desouz et al., 'Drugs affecting homocysteine metabolism: impact on cardiovascular risk', *Drugs*, vol. 62 (2002), pp. 605–16
45. J. B. Ubbink et al., 'The effect of a subnormal vitamin B-6 status on homocysteine metabolism', *Journal of Clinical Investigation*, vol. 98 (1996), pp. 177–84
46. R. Fijnheer et al., 'Homocysteine, methylenetetrahydrofolate reductase polymorphism, antiphospholipid antibodies, and thromboembolic events in systemic lupus erythematosus: a retrospective cohort study', *Journal of Rheumatology*, vol. 25 (1998), pp. 1737–42
47. M. Krogh Jense et al., 'Folate and homocysteine status and haemolysis in patients treated with sulphasalazine for arthritis', *Scandinavian Journal of Clinical Laboratory Investigation*, vol. 56 (1996), pp. 421–9
48. T. Apeland et al., 'Antiepileptic drugs as independent predictors of plasma total homocysteine levels', *Epilepsy Resident*, vol. 47 (2001), pp. 27–35
49. J. Vrbíková et al., 'Homocysteine and steroids levels in metformin treated women with polycystic ovary syndrome', *Experimental and Clinical Endocrinology and Diabetes*, vol. 110 (2002), pp. 74–6

Part 5

1. R. H. Allen et al., 'Metabolic abnormalities in cobalamin (vitamin B12) and folate deficiency', *FASEB Journal*, vol. 14 (1993), pp. 1344–53
2. M. F. Fenech et al., 'Folate vitamin B12, homocysteine status and chromosome damage in lymphocytes of older men', *Carcinogenesis*, vol. 18 (1997) pp. 1329–36
3. T. Shimakawa et al., 'Vitamin intake: A possible determinant of plasma homocysteine among middle-aged adults', *Annals of Epidemiology*, vol. 7, no. 4 (1997), pp. 285–93
4. O. Stanger, 'Physiology of folate in health and disease', *Current Drug Metabolism*, vol. 3 (2002), pp. 211–23
5. K. M. Koehler et al., 'Folate nutrition and older adults: challenges and opportunities', *Journal of the American Dietetic Association*, vol. 97 (1997), pp. 167–73
6. B. H. Patterson et al., 'Fruit and vegetables in the American diet: Data from the NHANES II survey', *American Journal of Public Health*, vol. 80 (1990), pp. 1443–9
7. L. B. Bailey, 'New reference intakes for folate: the debut of dietary folate equivalents', *Nutritional Review*, vol. 56 (1998), pp. 294–9
8. S. Hustad et al., 'Riboflavin as a determinant of plasma total homocysteine: effect modification by the MTHFR C677T polymorphism', *Clinical Chemistry*, vol. 46, no. 8 (2002), pp. 1065–1071

9. P. A. Ashfield-Watt et al., 'Methylenetetrahydrofolate reductase 677C–[TT & CT] genotype modulates homocysteine responses to a folate-rich diet or a low-dose folate supplement: a randomized controlled trial', *American Journal of Clinical Nutrition*, vol. 76 (2002), pp. 180–6

10. L. J. Riddell et al., 'Dietary strategies for lowering homocysteine concentrations', *American Journal of Clinical Nutrition*, vol. 71 (2000), pp. 1448–54

11. M. F. Fenech et al., *Carcinogenesis*, vol. 18 (1997), pp. 1329–36

12. H. O'Grady et al., 'Oral folate improves endothelial dysfunction in cigarette smokers', *Journal of Surgical Research*, vol. 106 (2002), p. 342

13. A. Cafolla et al., 'Effect of folate and vitamin C supplementation on folate status and homocysteine level: a randomised controlled trial in Italian smoker-blood donors', *Atherosclerosis*, vol. 163 (2002), pp. 105–11

14. E. P. Quinlivan et al., 'Importance of both folate and vitamin B12 in reduction of risk of vascular disease', *Lancet*, vol. 359 (2002), pp. 227–8

15. S. Friso et al., 'Low circulating vitamin B6 is associated with elevation of the inflammation marker c-reactive protein independently of plasma homocysteine levels', *Circulation*, vol. 12, no. 103(23) (2001), pp. 2788–91

16. K. Robinson et al., 'Hyperhomocysteinemia and low pyridoxal phosphate. Common and independent reversible risk factors for coronary artery disease', *Circulation*, vol. 92 (1995), pp. 2825–30

17. M. Leblanc et al., 'Folate and pyridoxal-5'-phosphate losses during high-efficiency hemodialysis in patients without hydrosoluble vitamin supplementation', *Journal of Renal Nutrition*, vol. 10 (2000), pp. 196–201

18. C. J. Bates et al., 'Plasma pyridoxal phosphate [P-5-P] and pyridoxic acid and their relationship to plasma homocysteine in a representative sample of British men and women aged 65 years and over', *British Journal of Nutrition*, vol. 81 (1999), pp. 191–201

19. K. Dalery et al. 'Homocysteine and coronary artery disease in French Canadian subjects: relation with vitamins B12, B6, pyridoxal phosphate, and folate', *American Journal of Cardiology*, vol. 75 (1995), pp. 1107–11

20. M. F. McCarty, 'High dose B6 in high stress patients (excess chronic stress associated with elevated homocysteine) for mood elevation, lowering of blood pressure and reduction of homocysteine', *Medical Hypotheses*, vol. 54 (2000), pp. 803–7

21. H. McNulty et al., 'Impaired functioning of thermolabile methylenetetrahydrofolate reductase is dependent on riboflavin status: implications for riboflavin requirements', *American Journal of Clinical Nutrition*, vol. 76, no. 2 (2002), pp. 436–41

22. S. Hustad et al., 'Riboflavin as a determinant of plasma total homocysteine: effect modification by the methylenetetrahydrofolate reductase C677T polymorphism', *Clinical Chemistry*, vol. 46 (2000), pp. 1065–71

23. H. Cass, 'SAMe – the master tuner supplement for the 21st century', published on www.naturallyhigh.co.uk

24. From a presentation by world authority Dr Reiter, at the American Anti-Aging Association's 11th International Symposium held in Las Vegas, December, 2002

25. L. A. Poirier et al., 'Blood determinations of S-adenosylmethionine, S-adenosylhomocysteine, and homocysteine: correlations with diet', *Cancer Epidemiology, Biomarkers and Prevention: A Publication of the American Association for Cancer Research*, vol. 10 (2001), pp. 649–55

26. D. O. McGregor et al., 'Betaine supplementation decreases post-methionine hyperhomocysteinemia in chronic renal failure', *Kidney International*, vol. 61, no. 3 (2002), pp. 1040–6

27. K. Brahmajee et al., 'Potential clinical and economic effects of homocyst(e)ine lowering', *Archives of Internal Medicine*, vol. 160 (2000), pp. 3406–12

28. A. J. Barak et al., 'Chronic ethanol consumption increases homocysteine accumulation in hepatocytes', *Alcohol*, vol. 25 (2001), pp. 77–81

29. M. H. Mar and S. H. Zeisel, 'Betaine in wine: answer to the French paradox?', *Medical Hypotheses*, vol. 53, no. 5 (1999), pp. 383–5

30. M. R. Malinow et al., 'Short-term folate supplementation induces variable and paradoxical changes in plasma homocyst(e)ine concentrations', *Lipids*, vol. 36 (2001), pp. S27–32

31. K. Koyama et al., 'Efficacy of methylcobalamin on lowering total homocysteine plasma concentrations in haemodialysis patients receiving high-dose folate supplementation', *Nephrology, Dialysis, Transplantation: Official Publication of the European Dialysis and Transplant Association*, vol. 17 (2002), pp. 916–22

32. R. Clarke et al., 'Lowering blood homocysteine with folate based supplements: meta-analysis of randomised trials, homocysteine lowering trialists' collaboration', *British Medical Journal*, vol. 316 (1998), pp. 894–8

33. 'B Vitamin Regimen May Protect Heart [after corrective heart surgery by lowering elevated homocysteine]', *Journal of the American Medical Association*, vol. 288 (2002), pp. 973–9

34. I. Kato et al., 'Epidemiologic correlates of serum folate and homocysteine levels among users and non-users of vitamin supplement', *Vitamin and Nutrient Research*, vol. 69 (1999), pp. 322–9

35. Garg, Rekha et al., 'Niacin treatment increases plasma homocysteine levels', *American Heart Journal*, vol. 138 (1999), pp. 1082–87

36. M. F. McCarty, 'Co-administration of equimolar doses of betaine [TMG] may alleviate the hepatotoxic risk associated with niacin therapy', *Medical Hypotheses*, vol. 55 (2000), pp. 189–94

37. A. Cafolla, F. Dragoni et al., 'Effect of folate and vitamin C supplementation on folate status and homocysteine level: a randomised controlled trial in Italian smoker blood donors', *Atherosclerosis*, vol. 163 (2002), pp. 105–11

Recommended reading

James Braly MD, *Dr Braly's Food Allergy and Nutrition Revolution,* McGraw-Hill/Contemporary Books, 1992

James Braly MD and Ron Hoggan, *Dangerous Grains,* Avery, 2002

James Braly MD with Jim Thompson, *Food Allergy Relief,* McGraw-Hill/Contemporary Books, 2000

Craig Cooney PhD and Bill Lawren, *Methyl Magic: Maximum Health Through Methylation,* Andrews McMeel, 1999

Jimmy Gutman MD and Stephen Schettini, *Glutathione GSH: Your Body's Most Powerful Healing Agent,* G&S Health Books, 2000

Patrick Holford, *The Optimum Nutrition Bible,* Piatkus, 1997

Patrick Holford, *Optimum Nutrition for the Mind,* Piatkus, 2003

Patrick Holford, *Say No to Heart Disease,* Piatkus, 1998

Patrick Holford, *Say No to Cancer,* Piatkus, 1999

Patrick Holford, *Say No to Arthritis,* Piatkus, 1999

Kilmer McCully, *The Homocysteine Revolution,* McGraw-Hill/Contemporary Books, 1999

Resources

www.thehfactor.com

Visit our website for up-to-date information on:

- How to test your homocysteine
- The latest research
- Homocysteine-lowering supplements and where to get them
- Success stories from others who've followed the H Factor programme
- Your questions answered
- Events and seminars
- Referral to a clinical nutritionist near you

Information for cardiac patients

UK

The British Cardiac Patients Association is a voluntary organisation offering help and information to cardiac patients and their carers via nationwide support groups and a bi-monthly publication *The Journal*. For details call 01954 202022, write to 2 Station Road, Swavesey, Cambridge CB4 5QJ, or visit the website at www.bcpa.co.uk.

Australia

The Heart Foundation of Australia website lists programmes and activities in your state or territory, as well as giving information about and contacts for local support groups. Visit the website at www.heartfoundation.com.au/links/index_fr.html.

New Zealand

The National Heart Foundation of New Zealand can give information about support groups across the country. Contact Catherine Robinson, MSc., the National Cardiac Rehabilitation Coordinator, at catheriner@nhf.org.nz, phone 09571 9191 ext. 798, or visit the website www.nhf.org.nz

South Africa

The Heart Foundation of South Africa publishes *Heart* magazine and has a network of support groups, Mended Hearts. Visit the website at www.heartfoundation.co.za/mendedhearts.asp

Homocysteine testing

The following laboratories offer homocysteine testing, available to the public:

YorkTest has developed the world's first homocysteine test that comes with a plasma separator, which means you can test yourself at home without the need to visit a laboratory to have a sample taken and the plasma separated by machine. The cost is £59.95, including postage. For details call 0800 074 6185 or visit www.yorktest.com.

Great Smokies Diagnostic Laboratory tests for homocysteine as part of a comprehensive cardiovascular risk assessment. Although this is usually only offered via a nutritional practitioner or healthcare professional, Great Smokies UK agent (Health Interlink) will send tests direct to members of the public. The cost for the complete test is £145. Call 01664 810 011 for details or visit www.health-interlink.co.uk.

Individual Wellbeing can do a homocysteine test at their London clinic or via mail order, but blood plasma needs separating out within half an hour of a blood sample being taken. The cost is £65. Call 020 7730 7010 for details or visit www.individual-wellbeing.co.uk.

If you visit your GP, they can get a homocysteine test run for you. However, they will rarely do this unless, for example, you have a high risk of heart disease.

For overseas readers, please go to www.thehfactordiet.com for an up-to-date list of laboratories offering homocysteine testing outside of the UK.

Meditation

UK
There are countless meditation approaches and courses available. Two that have received good feedback are the one-day Learn To Meditate courses offered by Siddha Yoga and courses at the London Buddhist Centre. Both groups have regional networks. For the Siddha Yoga Meditation Centre, write to 32 Cubitt Street, London WC1X 0LR, or call 020 7278 0035, and for the London Buddhist Centre, write to 51 Roman Road, London E2 0HU, or call 020 8981 1225.

Australia
The Siddha Yoga Foundation Australia holds training courses throughout Australia in capital cities and regional centres. Their website is www.siddhayoga.org.au.

New Zealand
As a first port of call, try www.nzhealth.net.nz/nzregister/meditate.html or www.buddhanet.net/nzbudir2.htm for listings

of meditation and Buddhist meditation centres and teachers throughout New Zealand.

South Africa
Buddhanet lists a variety of Buddhist meditation centres throughout South Africa: www.buddhanet.net/africame/africadir.htm.

Nutrition consultants

The Institute for Optimum Nutrition (ION) runs courses in nutrition from one-day workshops to a three-year Nutritional Therapy Diploma course that includes training in homocysteine management. They have a clinic, a list of nutrition practitioners across the UK, an information service and a quarterly journal – *Optimum Nutrition*. Contact ION at Blades Court, Deodar Rd, London SW15 2NU, call 020 8877 9993, or visit www.ion.ac.uk.

For a personal referral by Patrick Holford to a clinical nutritionist trained in homocysteine management in your area, please write to Holford and Associates, Carter's Yard, London SW18 4JR. Enclose your name, address, telephone number and brief details of your health issue. Alternatively, visit his website for an immediate referral at www.patrickholford.com.

Overseas readers can visit www.patrickholford.com for information about finding a nutritionist in their local area.

Natural Progesterone Information Society

NPIS provides women and their doctors with details on how to obtain natural progesterone information packs for the general public and health practitioners, and books, tapes and videos relating to natural hormone health. For an order form and prescribing details, please write with a stamped addressed envelope to NPIS, BCM Box 4315, London WC1N 3XX.

Directory of supplement companies

The following supplement companies offer supplements containing combinations of B6, B12, folate, B2, zinc, TMG and other homocysteine-reducing nutrients.

Higher Nature has a wide range of supplements including H Factors, which combines B6, B12, folic acid with TMG. Available by mail order. Call 01435 882880 for a full-colour catalogue and free newsletter, visit the website at www.higher-nature.co.uk, email sales@higher-nature.co.uk or write to them at Burwash Common, East Sussex TN19 7LX.

Solgar also sells a wide range of supplements including Homocysteine Modulators (with TMG, B6, B12 and folic acid). Available in good healthfood shops or by mail order by calling 01442 890 355. Or write to Solgar Vitamin and Herb, Adlbury, Tring, Herts HP23 5PT. For more information, visit www.solgar.com.

Health Interlink stock a product manufactured by Jarrow Formulas called Homocysteine PF (again with TMG, B6, B12, folic acid). Available by mail order. Call 01664 810 011 or visit www.health-interlink.co.uk for details.

Nutri sells a Homocysteine Redux formula (with B6, folic acid, B12 and TMG plus vitamin E, magnesium, zinc, selenium, N-acetyl-cysteine and others). This product is stocked in some healthfood shops or can be ordered via a nutritionist (see above for details). Call 0800 298 6280 for further information.

Australia
Solgar supplements are available in Australia. To find a shop stocking their products, visit www.solgar.com.au.

New Zealand
Solgar supplements are available in New Zealand. To find stockist information call 09 573 5101, email nzoffice@solgar.co.nz or visit www.solgar.com/International/newzealand.htm for further information.

South Africa
Bioharmony supplements are recommended. Bioharmony can be contacted at P O Box 18663, Wynberg, Cape Town, 7824, South Africa, Telephone: 021 797 8629, Facsimile: 021 797 8626, email: information@bioharmony.co.za, or visit: www.bioharmony.co.za.

Index

Page numbers in italics refer to diagrams

Other titles by Patrick Holford

£14.99 0 7499 2254 0

£16.99 0 7499 2213 3

£12.99 0 7499 1855 1

£5.99 0 7499 2336 9

£5.99 0 7499 2334 2

£6.99 0 7499 1920 5

NORADRENALIN — HAPPY & MOTIVATED

ADDS A METHYL GROUP IN ADRENAL GLAND

∴ MAKES ADRENALIN — ENERGY
                                AGRESSION

MELATONIN TO HELP YOU SLEEP

GLUTATHIONE — FLU OR COLD

BODY MAKE ABOVE